D0855742

LASER PIONEERS

Revised Edition

LASER PIONEERS

Revised Edition

Jeff Hecht

ACADEMIC PRESS, INC.
Harcourt Brace Jovanovich, Publishers
Boston San Diego New York
London Sydney Tokyo Toronto

This book is printed on acid-free paper. 

Portions originally published in *Lasers and Optronics* ®,
(formerly *Lasers and Applications*) a Gordon
Publications, Inc. publication.

Cover design by Elizabeth E. Tustian, text design by Catherine DiPasquale.

ACADEMIC PRESS, INC.
1250 Sixth Avenue, San Diego, CA 92101

United Kingdom Edition published by
ACADEMIC PRESS LIMITED
24-28 Oval Road, London NW1 7DX

Library of Congress Cataloging-in-Publication Data:

Hecht, Jeff.
Laser pioneers / Jeff Hecht. — Rev. ed.
p. cm.
Rev. ed. of: Laser pioneer interviews. 1985.
Includes bibliographical references (p.) and index.
ISBN 0-12-336030-7 (alk. paper)
1. Lasers—History. I. Hecht, Jeff. Laser pioneer interviews.
II. Title.
TA1677.H43 1992
621.36′6′09—dc20 91-17275
CIP

Printed in the United States of America
91 92 93 94 9 8 7 6 5 4 3 2 1

CONTENTS

Preface

As befits a book covering the rather convoluted history of laser development, *Laser Pioneers* itself has had a somewhat complex history. It began as a series of interviews in *Lasers & Applications*, a laser-industry trade magazine, published during 1985 to commemorate the twenty-fifth anniversary of operation of the first laser. The 13 original interviews and a brief overview of laser history were published as *Laser Pioneer Interviews* by High Tech Publications, then the publisher of *Lasers & Applications*. The magazine was later renamed *Laser & Optronics*, and sold to Gordon Publications Inc.

That first edition was small and never widely available. It represented considerable time and energy on the part of many people, however, and I wanted it to have broader distribution. I was delighted to find that Academic Press shared my feelings and was willing to publish an updated and expanded edition.

I have taken advantage of the intervening time to rewrite and greatly expand the overview of laser history, adding many new illustrations, including rare photographs taken at early laser conferences and in the Soviet Union. Most of the interviews have been expanded, with new reflections on the progress of laser technology.

One new interview has been added: Dennis Matthews, whose announcement of the first laboratory x-ray laser was published in the same year as the first edition. I also have expanded and updated both the bibliography of historic laser papers and the list of further readings. Unfortunately, Theodore Maiman declined to have his interview reprinted in this edition, so I have written an essay describing his work, based on published material including the original interview published in the May 1985 issue of *Lasers and Applications*.

Although my name appears on the title page, this book is not mine alone. First and foremost, it is the story of the 14 men who played important roles in one of the most dramatic scientific breakthroughs of the century: the laser. They graciously spent their time reflecting on their early work; they tolerated my questions and requests, and supplied valuable photographs.

Three other contributors to *Lasers & Applications* conducted some original interviews. C. Breck Hitz, then a contributing editor, interviewed Charles H. Townes, Arthur L. Schawlow, J. J. Ewing, and John M. J. Madey. Jim Cavuoto, then editor, interviewed Theodore Maiman. Richard Cunningham, who remains executive editor, interviewed William Bridges. They and Gordon Publications Inc., the present publisher of *Lasers & Optronics*, have graciously agreed to have their interviews published here.

Several others deserve special thanks for making this book possible. As publisher of *Lasers & Applications*, Carole Black arranged for publication of the original magazine interviews and the first edition of the book. Tom Farre supervised publication of the first edition. Louis Andreozzi of Gordon Publications Inc. helped with paperwork above and beyond the call of duty. Robert Kaplan of Academic Press supported the project despite agonizing delays.

These interviews were conducted in parallel with a more formal scholarly study, the Laser History Project coordinated by the American Institute of Physics. I have benefited from informal cooperation with the Laser History Project and with the project director, science historian Joan Lisa Bromberg. Her analysis of laser history was published as *The Laser In America: 1950–1970* by MIT Press in 1991. To aid future scholars, some material used in preparation of this book is being donated to the American Institute of Physics Center for the History of Physics, 335 East

45th St., New York, NY 10017-3483, tel. 212–661–9404. (The center plans to move to Washington, DC, in late 1992 or 1993.)

Jeff Hecht
Auburndale, Mass.
March, 1991

INTRODUCTION

An Overview of Laser History

The laser is three decades old, but its roots go back long before physicists had developed the theoretical concepts behind the laser or its microwave-emitting counterpart, the maser. The thread of ideas leading to the laser can be traced back to a theory of light emission proposed by Albert Einstein during World War I. However, the full implications of that work was not realized until much later. Einstein's ideas remained primarily of academic interest until the 1950s, when Charles H. Townes conceived of—and built—the maser.

Demonstration of the maser stimulated a wealth of ideas from physicists around the world, with the greatest concentration of work in the United States and a secondary concentration in the Soviet Union. Interest soon turned to extending maser concepts to shorter wavelengths, particularly to infrared and visible light, leading to the laser. Theoretical studies soon spawned a race to build the first laser, which eventually was won by Theodore H. Maiman on May 16, 1960. Many more lasers followed.

Maser and laser research led to several Nobel Prizes. The laser was a dramatic breakthrough, and it took some time to develop practical uses. One irreverent observer in the 1960s dubbed the laser "a solution looking for a problem," and there was some justice to that label early in the laser era. Now, however, lasers have found a host of important applications in fields ranging from medicine and research on atomic physics to home entertainment, fiber-optic communications, and military systems. Lasers help preserve vision, weld state-of-the-art razor blades, read product labels at supermarkets, and play music from compact discs.

As any scientific field matures, the pioneers look back and recall its origins. Many have interesting tales behind their discoveries. The stories of the laser pioneers are fascinating because many challenged conventional ideas about physics. If some of their breakthroughs seem simple today, it is only because we have the advantage of 20/20 hindsight. In their interviews, they give an inside view of the process of scientific innovation.

Laser history is replete with controversy as well as progress. The deepest dispute is a fundamental one: Who deserves credit for inventing the laser? Legal disputes over patents dragged on for years, with the final settlements made more than a quarter of a century after the first applications were filed. Court rulings may have settled the legal claims, but the arguments continue. This book can't resolve those disagreements, but it can let laser pioneers speak for themselves. This introduction gives an overview of laser history, explaining laser physics and putting the observations of the laser pioneers into context.

Stimulated Emission

The first step on the road to the laser was the publication of a paper by Albert Einstein (1916) describing how atoms could interact with light. He considered a photon, a quantum of light energy, which has the exact amount of energy needed to move an atom between two energy levels, as shown in Fig. 1.1. Physicists had previously thought only two things were possible. If the atom was in the lower energy state, it could absorb a photon and move to the higher energy level, increasing its internal energy. If the atom was in the higher energy level, it could release energy as a photon, and

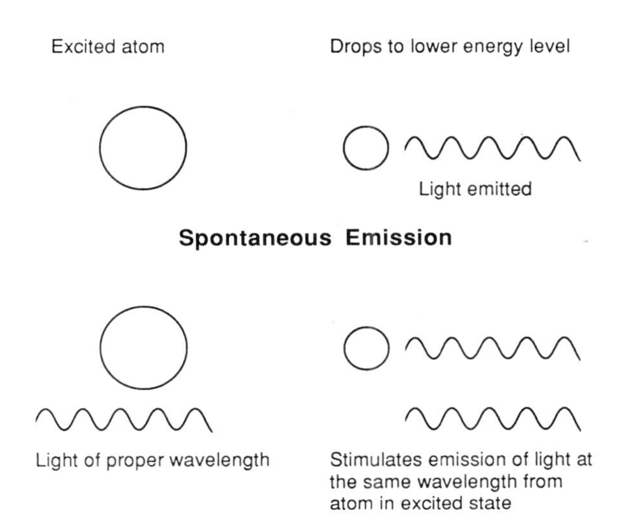

Figure 1.1 An excited atom (large circle) can emit light spontaneously to drop to a lower energy state (smaller circle), or it can be stimulated to emit light by light of the same wavelength. In a laser, the first spontaneously emitted light waves stimulate the emission of other light waves.

drop to the lower energy level, a process Einstein called "spontaneous emission" because it happens by itself. He also proposed a third alternative: If a photon arrived when the atom was in the higher energy level, it could stimulate the atom to emit another photon of the same energy, and drop to the lower energy state.

The importance of stimulated emission was far from obvious at first because it seemed to be an unusual event. Spontaneous emission can occur any time that atoms have excess energy that they can release as photons. Stimulated emission can occur only if light is present with precisely the right wavelength, so the photon energy equals the difference between the two energy levels. Spontaneous emission can produce some light at the right wavelength, but in practice it usually is absorbed by atoms in the lower energy state.

Absorption dominates as long as matter is in a condition called "thermodynamic equilibrium," long considered the normal state of the world in physics and chemistry classes. In thermodynamic equilibrium, physical and chemical processes have reached a steady state and are in balance. In that state, most atoms have a minimum energy. Some have higher energy, but the higher the energy of a state, the fewer atoms are in that energy level. Thus, there always are more atoms in the lower state of any pair of energy levels. If an atom drops from the higher to the lower level, releasing a photon with the transition energy, that photon is more likely to encounter an atom in the lower level than one in the higher level. Thus it is likely to be absorbed before it can stimulate emission from an atom in the higher-energy state, so stimulated emission is overwhelmed.

After World War I, German physicist Rudolf Walther Ladenburg (Fig. 1.2) became interested in spectroscopic theory and energy levels. He investigated atomic absorption, including dispersion, the change of a material's refractive index with wavelength. Absorption is unusually strong near "resonant" wavelengths, where the photon energy matches the transition from a common energy level to a higher one. Ladenburg's study of absorption near resonances led him to analyze an effect called negative dispersion,

Figure 1.2 Rudolf Walther Ladenburg (1882–1952), the German physicist who was the first to observe stimulated emission (courtesy of American Institute of Physics Neils Bohr Library).

a consequence of stimulated emission. His measurements (Ladenburg, 1928) were the first to confirm the existence of negative dispersion and stimulated emission, or what was then called "negative absorption." He continued his research for a few more years, measuring properties of neon gas after discharges were passed through it, but never raised discharge currents to levels high enough for "negative absorption" or stimulated emission to dominate.

Later observers, such as Arthur L. Schawlow, have suggested that physicists in the 1930s stopped short of producing conditions where stimulated emission would dominate because they were obsessed with thermal equilibrium, which they believed was the normal condition of matter throughout the universe. They apparently did not think they could move far enough away from equilibrium for the population of high-energy states to exceed that of low-energy states. That condition, called an "inverted population," is needed for stimulated emission to dominate and a laser to operate.

Another early conceptual barrier may have been the idea that a population inversion represented a "negative temperature." Elementary physics teaches that temperatures cannot go below absolute zero—0 K or −273° C—so "negative temperature" immediately sounds impossible. The term originated from the Boltzmann law for the population of energy levels at thermodynamic equilibrium, which depends on temperature. For positive temperatures, the law predicts that population decreases as energy level increases. If a population inversion exists, the formula implies that the temperature must be a negative number. The term survived to the dawn of the laser era, when Ali Javan titled an important study of populations in gases, "Possibility of production of negative temperature in gas discharges" (Javan, 1959).

It's intriguing to speculate the course of laser physics if someone in the 1930s had overcome these mental barriers to generate a population inversion. Indeed, it is possible that transient population inversions were produced, but were never detected because no one thought they could exist. However, the conceptual barriers were large enough that few ventured even theoretically into this seemingly forbidden territory.

One exception was Soviet physicist Valentin A. Fabrikant. In 1940 he wrote in his doctoral thesis that population inversion was necessary for "molecular amplification"—that is, domination by stimulated emission, or what we call laser action. He added, "Such a situation has not yet been observed in a discharge even though such a ratio of populations is in principle attainable . . . Under such conditions we would obtain a radiation output greater than the incident radiation and we could speak of a direct experimental demonstration of the existence of negative absorption" (Fabrikant, 1940, cited in Bertolotti, 1983).

Later in that decade, Willis E. Lamb, Jr., and R. C. Retherford came tantalizingly close to stumbling upon population inversions. In their studies of the fine structure of hydrogen—which led to Lamb's 1955 Nobel Prize in Physics—they briefly examined the populations of various energy levels in a discharge tube. They said that it was possible that an "induced emission could be detected," but they didn't try to do so themselves (Lamb and Retherford 1947, 1950).

The Maser

World War II triggered the development of radar technology, which helped open the microwave part of the spectrum. During the war, many scientists developed an interest in microwaves while working on radar. After the war, they were able to get military surplus microwave equipment to continue their research. This work led directly to the first device based on stimulated emission, the maser; the word was coined by Charles H. Townes, then at Columbia University, as an acronym for Microwave Amplification by the Stimulated Emission of Radiation.

The maser idea appears to have been conceived independently three times: by Townes at Columbia, by Joseph Weber at the University of Maryland, and by Alexander M. Prokhorov and Nikolai G. Basov at the Lebedev Physics Institute in Moscow. It was Townes, together with a postdoctoral assistant and a graduate student, who built the first maser.

Townes joined the Columbia physics faculty in 1948, after several years at Bell Laboratories, and became involved in microwave spectroscopy. His research at the Columbia Radiation Laboratory involved generating waves with millimeter wavelengths, shorter than the centimeter wavelengths of microwaves. That project advanced slowly, and by the spring of 1951 he was growing impatient. Townes recalls that the maser idea came to him while he was attending a scientific conference in Washington, DC. He woke early in the morning and left his hotel room to avoid disturbing his roommate, Arthur L. Schawlow, who then was a bachelor and accustomed to sleeping late. Sitting on a park bench, he mulled over the problems of generating millimeter wavelengths. He recalled,

> Perhaps it was the fresh morning air that made me suddenly see that this was possible: in a few minutes I sketched out and calculated requirements for a molecular-beam system to separate high-energy molecules from lower[-energy] ones and send them through a cavity which would contain the electromagnetic radiation [photons] to stimulate further emission from the molecules, thus providing feedback and continuous oscillation (Townes, 1978).

Townes took the idea back to Columbia, where he enlisted the help of postdoctoral fellow Herbert J. Zeiger and doctoral student James P. Gordon. They worked with a beam of ammonia molecules, seeking to produce a population inversion by isolating excited molecules. They directed the beam of excited molecules into a cavity resonant at the 24-gigahertz frequency of the ammonia transition. They thought that spontaneous emission from some excited ammonia molecules would stimulate others to emit at the same wavelength. They chose a resonant cavity to aid in coupling the emitted radiation to the excited ammonia molecules, and to reach higher levels of amplification.

After two years, Townes, Gordon, and Zeiger had spent about $30,000 of a Joint Services grant, but had not produced a maser. Some outsiders grew pessimistic, but the trio pressed on, and got the maser to work (Gordon, Zeiger, and Townes, 1954). Figure 1.3 shows Townes and Gordon with their second ammonia maser. It was only afterwards that the name "maser" was coined and quickly adopted by other researchers—although some cynics quipped that the acronym really stood for Means of Acquiring Support for Expensive Research (Bertolotti, 1983).

Townes was not the only one thinking of stimulated emission of microwaves. Weber analyzed prospects for obtaining stimulated emission from an inverted population shortly after joining the

Figure 1.3 Charles H. Townes (left) and James P. Gordon display the second ammonia maser they built at Columbia University (courtesy of Charles H. Townes).

faculty of the University of Maryland in 1951. No one has ever produced a population inversion in the way that he suggested, but he did realize that stimulated emission would be coherent, with the waves in phase. However, unlike Townes, Weber did not consider the possibility of making an oscillator, which would generate its own output. Instead, Weber envisioned a device that would amplify radiation from some other source. Weber's work was presented at a 1952 conference and published the next year (Weber, 1953).

Meanwhile, Prokhorov was leading a group of young physicists at Lebedev who were studying molecular spectroscopy. He and Basov sought to control the populations of various energy levels to enhance the sensitivity. This led to a detailed study, published in October 1954 (Basov and Prokhorov, 1954) of how to separate molecules with different energy in a molecular beam system, and how amplification could occur in a group of excited molecules. For his doctoral thesis Basov assembled the first Soviet maser, a few months after Townes' maser operated at Columbia. The two Soviet researchers' contributions were noteworthy, and they and Townes shared the 1964 Nobel Prize in Physics for developing the "maser–laser principle"; they are shown in Fig. 1.4 at about the time they received their Nobels. Figure 1.5 shows Prokhorov, Townes, and Basov in a laboratory at the Lebedev Physics Institute in Moscow at about the same time.

The field of maser research grew rapidly and involved many people who later pioneered laser development. One early advance

Figure 1.4 Nikolai G. Basov (1922–) and Alexander M. Prokhorov (1916–) of the P. N. Lebedev Physics Institute of the Soviet Academy of Sciences in Moscow shared the 1964 Nobel Prize in Physics with Townes for their work on masers and lasers (courtesy of AIP Meggers Gallery of Nobel Laureates).

Figure 1.5 Prokhorov, Townes, and Basov (left to right) at the Lebedev Physics Institute (courtesy of N. G. Basov).

was to move beyond the simple scheme Townes used in his molecular-beam ammonia maser. That system involved only two energy levels and required physical separation of excited molecules from those with lower energy. The population inversion occurred in a cavity where only excited molecules were present; it ended after stimulated emission dropped them to the lower level.

Sustained maser action requires a more complex scheme involving at least three energy levels, as shown in Fig. 1.6. The lowest level is the ground state, normally occupied by the molecules, which are excited to the highest level to produce a population inversion. Basov and Prokhorov (1955) proposed the first two schemes for three-level gas masers. One proposal was to excite molecules from the ground state to the highest energy level with an external radiation source. This would put more atoms in the highest state than in the intermediate level, generating a population inversion between those two upper levels, without the need to excite most of the molecules out of the lowest state. The other proposal was to remove molecules from the lowest energy level, reducing its population below the level in the slightly higher intermediate level. This would produce a population inversion between those low-lying states, leading to maser action.

Soon afterwards, Nicolaas Bloembergen, working at Harvard University, published the first proposal for a three-level solid-state maser (1956), which he describes in his interview. His proposal was more detailed and suggested some specific materials. Like the Soviets, he envisioned producing a population inversion by exciting the active species to a high energy level. Bloembergen's idea relied on Zeeman levels, states with energy proportional to an external magnetic field, which could be varied to change the level energy and hence the output wavelength. That was an important practical advantage, because early masers were limited to a very narrow range of wavelengths, corresponding to the precise energy of a molecular transition. Bloembergen's maser also maintained the extremely low noise levels of other masers. Another laser pioneer, Ali Javan, also worked on three-level masers. That concept was to prove important in laying the theoretical groundwork for the laser.

Although masers were soon overshadowed by lasers, they found some important applications in physical research, particularly for precision spectroscopy. At Harvard University, Norman F.

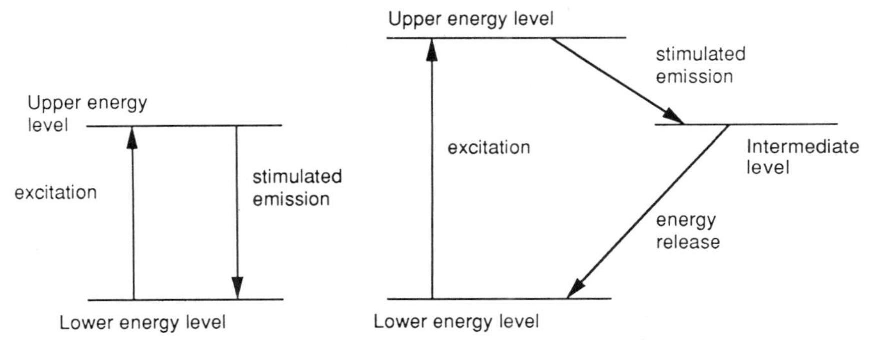

Figure 1.6 Only two energy levels are involved a two-level maser (left), such as the ammonia maser developed by Townes, Gordon, and Zeiger, who produced a population inversion by separating excited molecules from those in the lower state. The three-level maser (right) is more practical because the maser transition is not to the ground state; requirements for a population inversion are relaxed.

Ramsey, H. M. Goldenberg, and Daniel Kleppner used an atomic beam technique to make the first hydrogen maser in 1960. Starting with molecular hydrogen, they generated a beam of hydrogen atoms, then directed excited atoms into a microwave resonant cavity. The hydrogen maser can be made extremely stable, to better than one part in 10^{15}, making it extremely valuable for precision measurements (Ramsey, 1990). The hydrogen maser was among the achievements for which Ramsey shared in the 1989 Nobel Prize in Physics.

The Birth of Laser Theory

Even before masers began proliferating in the last half of the 1950s, a few physicists were looking at prospects for amplifying stimulated emission at wavelengths much shorter than microwaves. In the Soviet Union, V. A. Fabrikant and his students filed a patent application dated June 18, 1951, titled "A method for the amplification of electromagnetic radiation (ultraviolet, visible, infrared, and radio waves), distinguished by the fact that the amplified radiation is passed through a medium which, by means of auxiliary radiation or other means, generates excess concentration, in comparison with the equilibrium concentration of atoms, other particles, or systems at upper energy levels corresponding to excited states." However, that application is said not to have been accepted initially by the Soviet patent office and was not published until 1959 (Bertolotti, p. 115). It appears to have had little direct effect on laser research, even in the Soviet Union.

In the United States, Robert H. Dicke developed the concepts of superradiance and what he called the "optical bomb" (Dicke, 1954, 1964). His idea was to use a short excitation pulse to produce an inverted population, which would then generate an intense burst of spontaneous emission. He separately suggested that a pair of parallel mirrors, forming a Fabry–Perot interferometer, could serve as a resonant optical cavity in a patent titled "Molecular amplification and generation systems and methods," which he filed in 1956 and which was granted two years later (Dicke, 1958).

The first detailed proposal for building a laser—which at the time they called an "optical maser"—was published by Townes and Arthur L. Schawlow. Schawlow had been a postdoctoral fellow under Townes at Columbia until leaving to join Bell Labs in 1951, and they continued working together on a book on microwave spectroscopy, although not on masers. The two also maintained close personal ties—Schawlow married Townes' sister Aurelia and Townes consulted at Bell Labs. In 1957 both began thinking about the possibility of "infrared and optical masers," and after discussing the idea over lunch at Bell Labs, decided to collaborate. They spent several months working on the problem, as described in their interviews, which led to their famous paper "Infrared and Optical Masers," published in the December 1958 *Physical Review* (Schawlow and Townes, 1958).

That paper had a profound impact on American laser development. Preprints circulated at Bell Labs and Columbia before the journal came out, and formal publication was the starting gun for the great laser race that culminated in completion of the first laser. Some laser pioneers recall its impact in their interviews. Not everyone realized its importance, however. Bell Labs attorneys did not think the idea was worth patenting, and filed a patent application only after Townes insisted. That led to U. S. Patent No. 2,929,922, issued in 1960 (Schawlow and Townes, 1960).

Meanwhile, similar ideas were running through the mind of a 37-year-old Columbia graduate student, Gordon Gould. At the time, Gould was doing doctoral thesis research under Polykarp Kusch, who shared the 1955 Nobel Prize in Physics with Willis Lamb. Gould was hardly a typical establishment scientist of the placid 1950s. He and his first wife joined a Marxist study group in the mid-1940s. After the 1947 Soviet takeover of Czechoslovakia he left both the group and his wife, but that background haunted him during the anticommunist witch hunts of the 1950s. He had begun taking graduate courses at Columbia while teaching at the City College of New York, but lost that teaching job in 1954 after he refused to identify other members of the study group to a special committee of the New York board of higher education. That incident incensed Kusch, who secured a research assistantship so Gould could become a full-time graduate student.

Gould wrote down his laser ideas—including a definition of "laser" as Light Amplification by the Stimulated Emission of Radiation—in late 1957, and had them notarized by a candy store owner named Jack Gould (no relation) in what he hoped was the first step to getting a patent. He continued writing his ideas in notebooks, but realized that he would have to leave Columbia to work on the laser. That led to a series of misadventures, which Gould describes in his interview.

The path from maser to laser was far from obvious, because of the large physical differences between microwaves and visible light. Optical photon energies are thousands of times larger than those of microwaves, and the wavelengths are thousands of times less. This means that energy levels are much different, as are the materials used. One of the most important differences is in the type of resonant cavity used, and developing the cavities needed for laser oscillation proved one of the biggest challenges for laser developers.

Electromagnetic waves have resonances in cavities if the distance traveled through the cavity is an integral number of wavelengths. The simplest resonance is a cavity a half-wavelength wide, in which a round trip equals one wavelength, as shown in Fig. 1.7. Microwaves are measured in centimeters, and microwave cavities typically are on the order of a wavelength across, and enclosed on all sides. Visible light has wavelengths under a micrometer, so analogous cavities are obviously impractical.

The solution to that problem, recognized by Townes and Schawlow and by Gould, was an optical device known as a Fabry–Perot interferometer. The Fabry–Perot is simply two flat mirrors mounted parallel to each other, as shown in Figure 1.7, separated by many thousands of wavelengths. Light bounces back and forth between them, through the laser medium, stimulating the emission of more light. In practice, one mirror reflects all incident light, while the other transmits some light to form the laser beam.

The allocation of credit for "inventing" the laser remains controversial. Townes and Schawlow have been widely honored by the scientific community. Their *Physical Review* paper had a profound impact, and was the single biggest event triggering many research efforts that led to early lasers. Gould's notebooks and their offspring—his patent applications and proposals for research funding—had only minimal circulation, and essentially no impact on most of the scientific world.

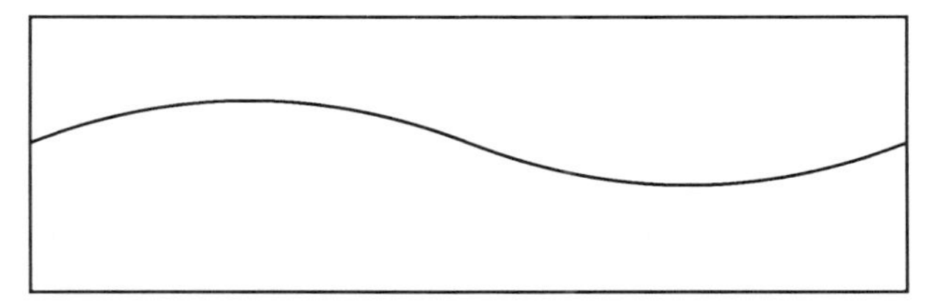

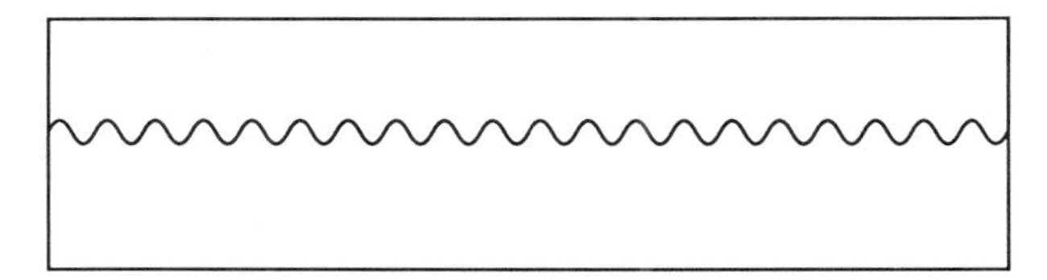

Figure 1.7 A microwave cavity is only a few wavelengths long (a), while a Fabry-Perot resonant optical cavity is many wavelengths long (b). Light waves actually are much smaller than shown; a typical helium-neon laser tube is hundreds of thousands of wavelengths long.

Gould has never achieved the professional eminence of Townes and Schawlow, both Nobel laureates in physics. By the early 1970s, he had become almost invisible in the laser world. However, Gould did make other contributions to lasers and fiber optics; although veteran researchers disagree on their importance, Gould cannot be lightly dismissed as a crackpot or plagiarist. When he finally began receiving patents in the late 1970s, some reports called him an underdog, an outsider who unfairly had been denied credit by the scientific establishment, while critics said his patents were undeserved. Ironically, the Townes–Gould controversy tends to overshadow the central fact that it was not one of them, but Theodore Maiman, who built the first laser.

Others were also exploring prospects for shorter-wavelength counterparts of the maser, including Basov and Prokhorov in the Soviet Union. The growing interest culminated in a race to translate theoretical concepts into practice and build the first laser.

The Great Laser Race

Publication of the Schawlow–Townes paper was the starting gun for the race to build what were then called optical masers. Some researchers, such as Townes, Theodore Maiman, and Nicolaas Bloembergen, had worked on microwave masers. Others, such as Peter P. Sorokin, Robert Hall, and C. Kumar N. Patel, came from other fields of physics after becoming intrigued with the laser concept. As time passed, more would enter the laser field as students, as did William Silfvast.

Early efforts concentrated on materials whose energy-level structures already were well known from spectroscopic studies. Ali Javan started working on the helium-neon gas laser at Bell Labs even before the Schawlow–Townes paper was published. At Columbia, Townes and two graduate students, Herman Z. Cummins and Isaac Abella, investigated a potassium-vapor scheme discussed in the *Physical Review* paper. In the Soviet Union, Basov studied semiconductors. With a few exceptions, such as Bell Labs, most of the research was modestly funded.

The largest research program was sponsored by the Advanced Research Projects Agency of the Department of Defense, the agency chartered to support risky research with high potential for rewards. The program had a peculiar history. As he developed his laser ideas, Gordon Gould realized that he could not continue pursuing both them and his graduate work. He left Columbia to work for a small company on Long Island, TRG Inc., and soon interested his employer in lasers. The company used Gould's ideas as the basis for a $300,000 research proposal to ARPA. Pentagon officials, dazzled by visions of laser weapons, were so excited that they gave TRG a contract for $1 million. Such increases are extremely rare.

Gould, like Townes, initially concentrated on alkali-metal vapors. The generous Pentagon funding let TRG investigate many laser candidates, but the program was ill-fated from the beginning, as Gould describes in his interview. Security restrictions came with the military money, and that caused problems for Gould. Although the worst anticommunist hysteria had passed, the Marxist skeleton in Gould's closet was enough to prevent him from getting a security clearance. TRG scientists trying to build lasers could consult with Gould, but they could not tell him details about classified research. Moreover, alkali-metal vapors would prove to be very difficult to make into lasers.

Most of the entrants in the great laser race gathered at the Schwang Lodge in the Catskill Mountains of New York for the first International Quantum Electronics Conference, September 14–16, 1959. It was a time of excitement remembered by all the participants, and attracted Basov, Prokhorov, and a few other Soviet scientists. Two photos from that conference are shown in Fig. 1.8.

The Ruby Laser

Among the maser materials considered for use in lasers was synthetic ruby, aluminum oxide doped with chromium atoms. The chromium lines were useful in masers, and their spectroscopy was well known. At Bell Labs, Schawlow considered ruby as a laser material, but in 1959 he publicly dismissed it as unsuitable. That opinion was based on inadequate data, and it was soon proved wrong.

Meanwhile, Theodore Maiman was trying to use his knowledge of ruby masers to make a laser at Hughes Research Laboratories in Malibu, California. He began working with ruby because it was well known; at first he thought he could switch to a better material later, when he more fully understood laser requirements. However, he eventually convinced himself that Schawlow was wrong, and that ruby would make a good laser. As related in the chapter describing his work, he forged ahead, while Hughes management grew skeptical. By the time he succeeded in making the ruby laser work for the first time, on May 16, 1960, he was not supposed to be working on the program.

Maiman's laser was small and elegant: a ruby rod, with its ends silvered to reflect light, which he placed inside a spring-shaped flashlamp. His success is undisputed, but he almost immediately ran into problems reporting it. Hughes management reacted enthusiastically once the laser worked, and sponsored a full-fledged press announcement in early July. However, the public-relations photographer commissioned to immortalize the first laser on film wasn't satisfied with it. He thought the device was too small, and

Figure 1.8 The first International Quantum Electronics Conference attracted most leading figures in laser research. The group photo (a) shows (from left) James P. Gordon, Nikolai Basov, Herbert Zeiger, Alexander Prokhorov, and Charles Townes (courtesy of C. H. Townes). The other photo shows Ali Javan talking with Basov (courtesy of N. G. Basov).

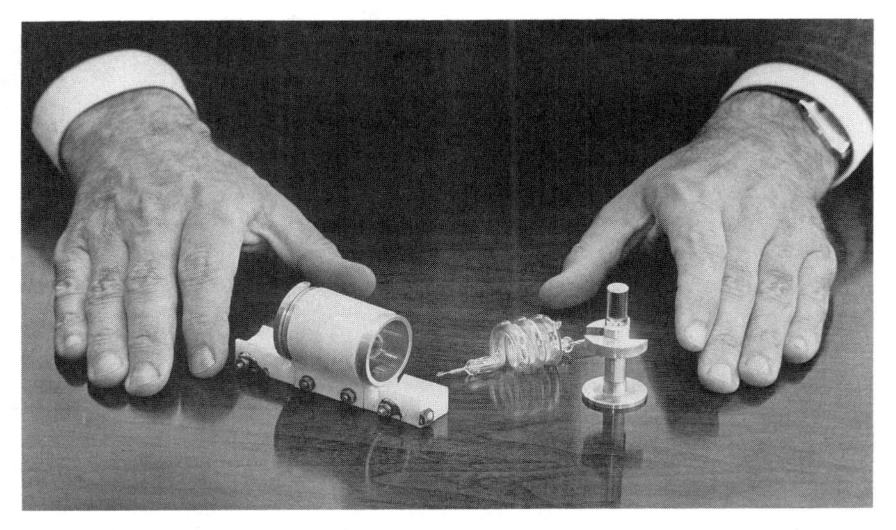

Figure 1.9 Theodore H. Maiman's first laser was small enough to fit into the palm of a hand; it is shown here dissembled (courtesy of Corning Inc).

insisted that Maiman pose with a bigger flashlamp and ruby rod. Today Hughes is still distributing those pictures, showing Maiman with what isn't really the first ruby laser. That first laser, small enough to be held in one hand, is shown in Fig. 1.9.

A more serious problem came when Maiman submitted his paper for publication. The then-new *Physical Review Letters* summarily rejected it as "just another maser paper." The journal's founding editor Samuel Goudsmit, a theoretician best known as the codiscover of electron spin, had grown tired of the glut of maser papers arriving at his office, and decided that they no longer merited rapid publication in his journal. Maiman hurriedly prepared a concise 300-word report, which was immediately accepted by the British weekly *Nature.* When efforts to convince Goudsmit of his error failed, the *Nature* paper, published August 6, 1960, became the first report of a working laser (Maiman, 1960a). Maiman later published a more detailed analysis in *Physical Review* (Maiman, 1961; Maiman *et al.*, 1961).

In their interviews, some laser pioneers recall when and how they heard the news of Maiman's laser. It was not long before other laboratories had made their own ruby lasers, although some used the flashlamp shown in the press release, rather than the one Maiman actually used. Schawlow's group at Bell Labs was among

Figure 1.10 Arthur L. Schawlow (left) adjusts an early ruby laser at AT&T Bell Laboratories in a 1960 demonstration, as C. G. B. Garrett readies a camera to record the laser pulse (courtesy of AT&T Bell Laboratories).

the first to get one working (Collins *et al.*, 1960). Their laser, shown in Fig. 1.10, was considerably larger than Maiman's. Soon afterwards, laser action on slightly different lines in "dark" or "red" ruby, which has a higher concentration of chromium ions than in the "pink" ruby used by Maiman, was reported by Bell Labs (Schawlow and Devlin, 1961) and another group (Wieder and Sarles, 1961) in *Physical Review Letters*.

The unusual sequence of publications has confused a few observers, especially those who looked only at American journals. Maiman did publish a paper on optical emission from ruby in *Physical Review Letters* (Maiman, 1960b), but it did not report laser action. Schawlow's group was the first to report a ruby laser in *Physical Review Letters* (Collins *et al.*, 1960), but they only replicated Maiman's demonstration. Some overenthusiastic statements from Bell Labs, and the notion that theorists deserve credit for "inventing" the laser, have muddied the picture further. However, it is clear that Maiman, not Bell Labs, deserves the credit for making the first laser.

The Great Laser Boom

Virtually everyone in the great laser race but Maiman had concentrated on complex schemes to generate an inverted population and strengthen what they expected to be a weak signal. Maiman's laser was so elegant in its simplicity, and produced such unexpectedly powerful pulses, that it made many researchers rethink their approaches. Peter P. Sorokin and Mirek Stevenson of the IBM T. J. Watson Research Center in Yorktown Heights, New York, switched to a flashlamp-pumped rod design from a more elaborate approach involving a polished block of laser crystal after learning of Maiman's results. The first time they flashlamp-pumped a cryogenically cooled crystal of uranium-doped calcium fluoride, in November 1960, it lased, and they had the second type of laser (Sorokin and Stevenson, 1960), as Sorokin describes in his interview.

Sorokin's uranium laser emitted infrared light at 2.5 micrometers, and because of factors including the need for cryogenic cooling it has never found practical applications. However, it marked another important milestone: It was the first laser to involve transitions among four energy levels, as shown in Fig. 1.11. The distinction between three- and four-level lasers is as significant as that between two- and three-level masers. While three-level lasers work, four-level lasers are more powerful and versatile systems. Four-level lasers can generate steady beams, while three-level systems are limited to pulsed operation. In short, four-level lasers work better.

Three-level laser (e.g. ruby) terminates in the ground state making a population inversion difficult and limiting it to pulsed operation

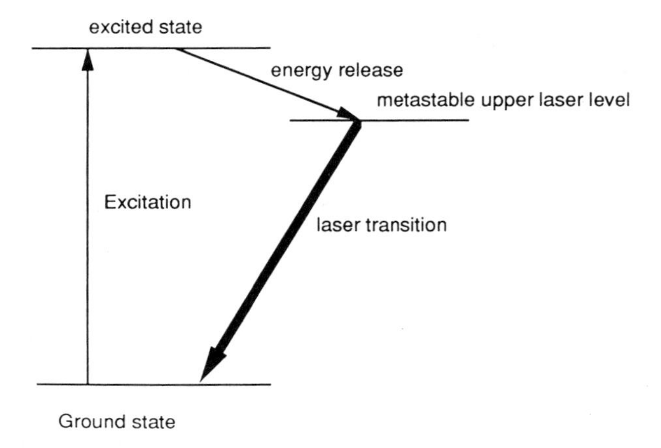

Four-level laser (e.g. neodymium) terminates above the ground state in a level which normally is unpopulated, allowing continuous operation

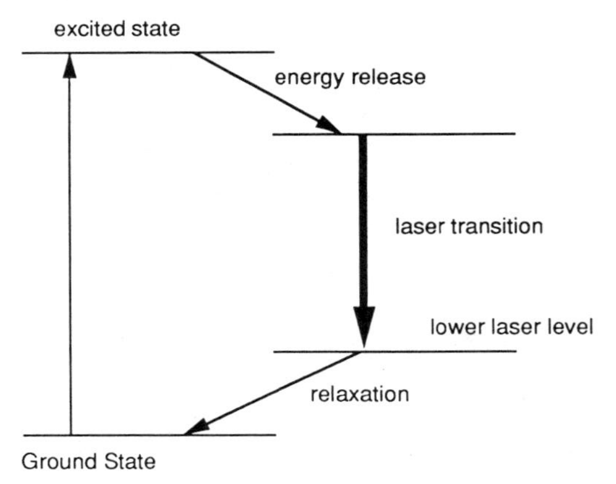

Figure 1.11 In three-level lasers such as ruby, more than half the laser species must be excited from the ground state to produce a population inversion, so operation normally is limited to pulses. Four-level lasers are more efficient because the lower laser level normally is unpopulated.

At room temperature, ruby is a three-level laser (although at low temperatures it becomes a four-level laser). Light from a flashlamp excites chromium atoms in the ruby to a high energy level, from which they quickly drop to the upper level of the laser transition. That upper laser level is called "metastable" because it has an unusually long lifetime, so many ruby atoms accumulate there, producing a population inversion. Emitting laser light drops the atoms back down to the ground state. The problem with a three-level system is that the population inversion needed for laser action is possible only if most chromium atoms in the crystal can be excited out of the ground state. This had seemed a formidable problem to many early developers, but Maiman solved it by using short, intense pulses from a flashlamp.

The four-level approach avoids the need to depopulate the ground state. As in ruby, the atoms start in the ground state, and are excited to a high-level state from which they decay to a metastable state that is the upper laser level. However, the lower laser level is not the ground state but rather a level with enough energy to be largely unpopulated before laser operation begins. The smaller the population of the lower laser level, the fewer atoms need be excited to produce a population inversion. As long as atoms or molecules leave the lower laser level, its population remains low, and the laser can emit a continuous beam. (Not all four-level lasers, however, can emit continuously.)

Sorokin and Stevenson also demonstrated a second four-level solid-state laser, based on divalent samarium ions (Sm^{2+}) in calcium fluoride. They operated the samarium laser at 708.5 nanometers at the edge of the red spectral region (Sorokin and Stevenson, 1961) a few weeks after the uranium laser. Like the uranium laser, the samarium laser requires cryogenic cooling and has yet to find practical applications.

The First Gas Laser

The proliferation of solid-state lasers was somewhat unexpected. Most early proposals were for gas lasers, including vapors of the alkali metals in Group 1a of the periodic table: sodium, potassium, and cesium. Just before the end of 1960, however, the well-funded gas laser program at Bell Labs finally reported a triumph. Ali Javan, William R. Bennett, Jr., and Donald R. Herriott at the Murray Hill, New Jersey lab made the first helium-neon laser (Javan, Bennett, and Herriott, 1961).

Javan studied under Townes at Columbia, and became interested in optical lasers at about the time he moved to Bell Labs. He decided that a gas discharge in a mixture of helium and neon would be a particularly clean system. He first demonstrated amplification, but building a laser able to oscillate was not an easy task because he lacked the tools available today. The mirrors had to be aligned precisely parallel to each other to get the helium-neon laser to oscillate, but that alignment was difficult to perform without a continuous-output laser. After painstaking effort, Javan, Bennett, and Herriott finally got the laser to work at 4:20 PM on December 12, 1960.

Their helium-neon laser was not the red type widely used today in applications ranging from reading product codes at supermarket checkout counters to aligning construction equipment. It emitted in the near infrared at a 1.15-micrometer transition, which Javan had calculated was stronger than the now-familiar red line. Javan says in his interview that he could not have made a red laser then, because its gain is high only in tubes with bore diameters so small that mirror alignment is very difficult.

The first operation of a helium-neon laser on the 632.8-nanometer red line did not come until more than a year later, when A. D. White and J. Dane Rigden (1962) demonstrated the red helium-neon laser at Bell Labs (Fig. 1.12). By that time Javan had moved to the Massachusetts Institute of Technology, where he remains on the faculty.

The success of the red "He-Ne" laser has proved that Javan's belief in the helium-neon system was well-founded. For more than 20 years, the red helium-neon laser was the most common type of laser. In volume, inexpensive semiconductor lasers took over first place in the mid-1980s, as sales of compact disc players and laser printers boomed. However, hundreds of thousands of red helium-neon lasers are still being sold each year, and manufacturers have developed commercial helium-neon lasers that emit on weaker visible lines. The helium-neon laser remains the most common gas laser.

The advent of the helium-neon laser marked two crucial milestones. It was the first laser to emit a continuous beam, preferable for many applications to the pulses from ruby and the solid-state lasers demonstrated by Sorokin and Stevenson. This was possible because neon, the light-emitting species, is a four-level laser. The

Figure 1.12 Dane Rigden looks at the visible beam from the red helium-neon laser that he and Alan White developed at Bell Labs in 1962 (courtesy of AT&T Bell Labs).

helium-neon laser also was the first gas laser. It showed not only that laser action was possible in a gas, but that it could be excited by an electric discharge and that it could rely on energy transfer between two different species—the helium atoms captured energy from electrons and transferred it to neon atoms. It is easier to put new gases in a laser tube than to grow new crystals, so gas lasers would soon outnumber solid-state lasers.

More New Lasers

The second gas laser is a fascinating footnote to laser history, but was never a practical device. Working under the Pentagon contract with TRG, Paul Rabinowitz, Steve Jacobs, and Gould (1962) succeeded in obtaining laser action from cesium vapor at 3.2 and 7.2 micrometers in the infrared. Instead of exciting the vapor with an electric discharge, they excited the cesium atoms with ultraviolet light at 388.9 nanometers from helium. As Gould relates in his interview, the TRG group exhaustively studied many laser candidates before finally getting one to work.

The system finally demonstrated at TRG was similar to schemes proposed both by Gould and by Schawlow and Townes. All sought to create a population inversion in an alkali-metal vapor by exciting the atoms with light at a wavelength close to that of spectral lines in the metal vapor. However, few such systems have worked, and none proved practical.

The other new laser demonstrated in 1961 was a solid-state type that would have much more practical impact. L. F. Johnson and K. Nassau (1961) at Bell Labs demonstrated laser action at 1.06 micrometers from neodymium ions in a host crystal of calcium tungstate. Neodymium is a four-level laser, and before long it was operated continuously at room temperature—an important breakthrough for solid-state lasers. Later that year Elias Snitzer (1961) at the American Optical Corporation showed that the neodymium ion could operate as a laser in optical glass.

The attractions of neodymium were clear enough that other researchers soon began studying its laser performance in different host materials. Laser action in the most successful of these, yttrium aluminum garnet (YAG) was first demonstrated in 1964 at Bell Labs (J. E. Geusic, H. M. Marcos, and L. G. Van Uitert, 1964).

The First Commercial Lasers

It didn't take long for laboratories around the world to duplicate Maiman's ruby laser—although many groups used the large flashlamp showed in the Hughes press-release photo, not the small one that Maiman had actually used. Not everyone, however, wanted to try building his own. The first commercial lasers began appearing in about 1961—large ruby types designed to deliver high-power pulses. Some were built by large corporations that had been doing laser research; others came from small firms started by laser researchers.

Maiman was one of the first to start building commercial lasers. He left Hughes shortly after reporting the ruby laser to head a group developing commercial lasers at a short-lived company called Quantatron in Santa Monica, California. When Quantatron ran out of money, Maiman took the laser group to form the core of Korad Inc., also in Santa Monica. Union Carbide funded the new company under an agreement that gave it the right to acquire control of Korad in five years. Korad soon became a major force in the laser market.

Other companies also built ruby lasers. One of the first was Trion in Michigan, which later was acquired by Lear Seigler and renamed the Lear Seigler Laser Systems Center. American Optical got an early start making neodymium-glass lasers, but never effectively capitalized on its work. It eventually spun off its laser group as an independent company, Laser Inc. in Sturbridge, Massachusetts, which later was acquired by Coherent Inc., and became part of a partnership between General Electric and Coherent. Only one early maker of ruby lasers, Raytheon Inc., still makes commercial solid-state lasers, although Hughes Aircraft makes military lasers.

Meanwhile, other companies started making gas lasers, initially concentrating on helium-neon. One was Spectra-Physics Inc., founded in 1961 by three men from Varian Associates, Herbert Dwight, Robert Rempel, and Earl Bell. Spectra-Physics developed a commercial laser as a joint venture with Perkin-Elmer, and went on to become the world's biggest manufacturer of lasers (although it sold its helium-neon laser business in the late 1980s). Another early maker of helium-neon lasers, founded in 1960 by fiber-optics pioneer Narinder Kapany, was Optics Technology Inc., which went out of business in 1973. RCA was one of the first large companies to make helium-neon lasers, but it left the business in the 1970s, many years before it was gobbled up by General Electric Company.

Laser-Related Developments

As early lasers became available in more laboratories, new discoveries were made, both about their properties and their potential applications. Theorists A. G. Fox and Tingye Li of Bell Labs laid valuable groundwork when they published an extensive theoretical analysis of oscillation patterns inside laser resonators (Fox and Li, 1961). Although their study was completed before anyone had operated a continuous laser, it remains a definitive analysis of laser oscillation.

At about the same time, Peter A. Franken, A. E. Hill, C. W. Peters, and G. Weinreich of the University of Michigan discovered that powerful pulses from a ruby laser changed as they passed through a quartz crystal (Franken *et al.*, 1961). Some light was converted from the 693-nanometer red ruby wavelength to 347 nanometers in the ultraviolet. The frequency of that ultraviolet light is twice the ruby laser frequency. Such second harmonic generation is one of many nonlinear optical phenomena, which can change the wavelength or frequency of light. This discovery stimulated the development of nonlinear optics and spectroscopy, research for which Nicolaas Bloembergen and Arthur Schawlow shared the 1981 Nobel Prize in Physics.

Another important development was a technique for changing the characteristics of a resonant laser cavity to generate short, powerful pulses. The concept, called Q-switching because it changes the cavity "quality factor," suppresses gain in the resonator so power can be stored in the laser medium, then quickly increases the cavity quality factor so the laser emits the accumulated energy in a pulse lasting on the order of 10 nanoseconds (10^{-8} second). It was first demonstrated by Robert W. Hellwarth and R. J. McClung at Hughes Research Laboratories (McClung and Hellwarth, 1962).

These developments in turn helped to enhance the usefulness of lasers. The study of laser oscillation led to improvements in resonator structure, particularly important for continuous lasers with low gain, such as helium-neon. Nonlinear optics and harmonic generation made new wavelengths available. Q-switching made available higher peak powers and shorter pulses—developments that opened new areas for research and led to practical applications, such as hole-drilling and measuring the range to military targets.

The Semiconductor Laser

The next major development in laser physics was the semiconductor laser. Invention of the transistor had made semiconductor physics a hot topic in the 1950s and probably helped speed development of the semiconductor laser. Light emission in semiconductors, however, had already been known for half a century. In 1907 H. J. Round in England found that application of an electric field causes silicon carbide (a semiconductor) to emit light. That "electroluminescent" emission comes from the junction between parts of the semiconductor with different concentrations of the impurities that carry electric current.

In modern electronic terms, this emitting zone is *p-n* junction, the boundary between regions where current is carried by holes (*p*-type semiconductor) and by electrons (*n*-type). If a voltage is applied so that both carriers flow toward the junction, the two recombine at the boundary and release energy. Certain semiconductors release some of that energy as light. This phenomenon is the basis of the light-emitting diode, or LED.

Given the wide interest in semiconductors and lasers, in retrospect it is not surprising that a number of people considered producing a population inversion and maser or laser action in semiconductors. The famous mathematician John von Neumann on September 16, 1953, sent Edward Teller a manuscript outlining his ideas for light amplification by stimulated emission in semiconductors (Dupuis, 1987b; von Neumann, 1987). He apparently had the idea of using a *p-n* junction in the semiconductor, although he did not use the term (Dupuis, 1987a). Von Neumann's notes were not published until long after his death, however, and had no real impact on semiconductor laser research.

Two Japanese researchers, Yasushi Watanabe and Jun-ichi Nishizawa, independently thought of generating stimulated emission across a semiconductor junction. In 1957 they applied for a patent on a "semiconductor maser," and on September 20, 1960, they received Japanese patent 273217 (Bertolotti, p. 165). They proposed enclosing the resonant cavity completely, as in microwave oscillators, and suggested it should be possible to obtain 4-micrometer emission from tellurium. Their patent application apparently had little impact elsewhere, however.

Pierre Aigrain of the Ecole Normale Supérieure discussed the possibility of making semiconductor lasers as far back as 1956, and in June 1958 presented a paper on the idea at a conference in Brussels. His paper was never published, however (Dupuis, 1987a, p. 652).

Basov's group in Moscow did the most extensive theoretical work and made a number of proposals, including the approach that proved successful: using *p-n* junctions in highly doped semiconductors (Basov, Kronkhin, and Popov, 1961). Others who analyzed prospects for stimulated emission from semiconductors included Maurice G. A. Bernard and G. Duraffourg (1961), and C. Benoit à la Guillaume (1961) in France. These theoretical studies, however, failed to identify some key features of semiconductor lasers, and as Russell Dupuis observed (1987a, p. 654), "[T]here was no clear understanding of what to expect experimentally when laser operation was achieved."

The theoretical papers did stimulate experiments that yielded encouraging results. A Leningrad group reported a slight decrease in the bandwidth of emission from cryogenically cooled gallium arsenide diodes at high drive currents (Nasledov *et al.*, 1962), a sign that stimulated emission might be occurring. At the Solid State Device Research Conference in July 1962, R. J. Keyes and T. M. Quist of MIT Lincoln Laboratories reported making GaAs diodes, which emitted incoherent light with internal quantum efficiency they estimated at 85% (Keyes and Quist, 1962). Other reports at that conference generated a level of excitement that Robert Hall describes in his interview.

In retrospect, it seems that most groundwork had already been laid for the semiconductor laser, and the July conference marked the start of a race to build one. It almost ended in a photo finish. Hall attended the meeting and took his ideas for making a semiconductor laser home to General Electric's Research and Development Laboratories in Schenectady. By September, he had a working diode laser (Hall *et al.*, 1962). Marshall I. Nathan's group at IBM Watson Research Center in Yorktown Heights, New York, followed within a matter of days, with a paper received on October 4 at *Applied Physics Letters* (Nathan *et al.*, 1962). Later that month researchers at Lincoln Labs (Quist *et al.*, 1962) and Nick Holonyak at the General Electric Laboratory in Syracuse (Holonyak and Bevacqua, 1962) also reported semiconductor lasers.

The four independent groups had come up with remarkably similar devices. The first three used *p-n* junctions in GaAs, which were cooled to 77 K, the temperature of liquid nitrogen, and driven by high-current pulses lasting about a microsecond. Their output was at 840 nanometers in the near infrared. Holonyak's diode was made of gallium arsenide-phosphide, so it emitted at shorter wavelengths of 600 to 700 nanometers. All but the IBM group polished the ends of the semiconductor crystal to serve as mirrors that provided feedback for laser oscillation in a resonant cavity.

Ironically, GTE Laboratories may have come up with the ideas needed for a working semiconductor laser in late 1961 but failed to move as rapidly as the other groups. Sumner Mayburg of GTE labs reported efficient light generation from *p-n* junctions in GaAs at an American Physical Society meeting in March 1962, but his postdeadline paper was never published. GTE researchers apparently documented the semiconductor laser concept in December 1961 but never published their proposal, and did not operate a laser until November 1962 (Dupuis, 1987a, note 19).

These first semiconductor lasers were only a beginning, and it would take years for more practical technology to evolve. Early experiments showed that carrier recombination at a junction produced excited states that could generate stimulated emission, as well as the spontaneous emission from LEDs. Early devices, however, operated as lasers only at extremely high current densities: on the order of 10,000 amperes per square centimeter of junction area. They could operate only in pulsed mode, and only far below room temperature, or the high drive currents would burn them out.

These high currents reflected an inherent limitation of early semiconductor laser designs: poor confinement of the drive current and the laser emission. The first lasers were "homojunction" types, in which the light-emitting active layer was bounded on top and bottom by the same material. Confinement was improved—and threshold current reduced—when developers began using more complex structures. The key development that allowed operation of a semiconductor laser at room temperature was the "double heterojunction" or "double heterostructure" laser, in which the active layer is sandwiched between two layers of a slightly different material. Both light and current are confined in the active layer by using a material in which the "band gap"—the energy difference between electrons that conduct current and electrons that form crystalline bonds—is larger than that in the active layer. The first such device was made in 1968 (Kasonocky *et al.*, 1968). Later groups at Bell Labs (Hayashi *et al.*, 1970) and in the Soviet

Union (Alferov *et al.*, 1970) made the first semiconductor lasers capable of continuous operation at room temperature. Izuo Hayashi of Bell Labs is shown with one laser in Fig. 1.13; another key member of the team, Morton Panish, is shown in Fig. 1.14.

The first Bell Labs devices lasted for only a few hours at room temperature, but lifetimes steadily improved. By 1975, semiconductor lasers able to emit continuously reached the commercial market, initially at a few thousand dollars each. Steady technical advances helped trigger a tremendous expansion in the semiconductor laser market that continues today. New structures confine the active layer not merely within a thin layer, but within a thin stripe, further reducing threshold currents and improving the quality of output and device lifetime. New material systems have

Figure 1.13 Izuo Hayashi points to a room-temperature diode laser mounted on a heat sink. He was part of the Bell Labs group that made continuous-wave, room-temperature diode lasers in 1970 (courtesy of AT&T Bell Labs).

Figure 1.14 Morton B. Panish, head of the Bell Labs group that developed room-temperature diode lasers (courtesy of AT&T Bell Labs).

been developed for semiconductor lasers emitting at both longer and shorter wavelengths than the 750 to 900 nanometer range of GaAlAs. Mass-production techniques allow high-quality diode lasers to be sold in quantity for just a few dollars each, for applications such as compact disc players and laser printers. Sales of diode lasers passed the million mark in 1985, and by 1988 had passed 20 million a year.

The Gas Laser Boom

Early gas lasers had limited output power. Bell Labs built a monstrous helium-neon laser, but it generated only 150 milliwatts. C. Kumar N. Patel, then a young physicist at Bell Labs, realized that laser action from atomic gases was limited in power, and decided to try a different approach: molecular gases.

Carbon dioxide was his first choice, and after calculations indicated it should work well, he tried exciting the pure gas in an electrical discharge. On the first shot in 1963, he obtained tens of milliwatts at 10.6 micrometers in the infrared. His carbon-dioxide laser (Patel, 1964) was the first high-power gas laser. Soon after he demonstrated the carbon-dioxide laser, he realized that he had been too quick to dismiss molecules containing only two atoms. He went back to try carbon monoxide, which also worked. Patel and others gradually refined the carbon-dioxide laser, first adding nitrogen to transfer the discharge excitation more efficiently to the carbon-dioxide molecules, then adding helium or water to de-excite the lower laser level in carbon dioxide and thereby increase laser output power. That work also led to other infrared molecular gas lasers that operated on similar principles.

The next major type of gas laser to be discovered was the ion laser, in which the active species was an ion missing one or more electrons. The first of these was a mercury-ion laser discovered in late 1963 by Earl Bell and Arnold Bloom at Spectra-Physics Inc. (Bell, 1964). They observed four lines, two in the visible and two in the infrared, from a pulsed gas discharge in a tube containing a 500:1 mixture of helium and mercury, with peak power to 40 watts.

The helium-mercury laser never became an important commercial type, but efforts to duplicate and understand it led William Bridges to discover the argon-ion laser, which has become important. As Bridges explains in his interview, the discovery was an accidental one that came from his efforts to understand how the buffer gas affected mercury-laser operation. He demonstrated a neon-mercury laser, but ran into problems with an argon-mercury version. To check mirror alignment, he pumped out the tube and refilled it with a helium-mercury mixture. It worked—and also produced a previously unobserved blue-green line at 488 nanometers (nm). Careful checking showed that the 488-nm emission came from argon left in the tube. Bridges then tried pure argon in a clean tube, and observed laser action on visible lines of singly ionized argon (Bridges, 1964), The same lines were discovered independently by two other groups: by Guy Convert's group at CSF in France, and by William Bennett and some of his graduate students at Yale University, where he had returned after spending a year at Bell Labs.

Bridges's work touched off a flurry of research on laser action in rare-gas ions, leading to demonstrations of krypton, xenon, and neon-ion lasers. Those early lasers all were pulsed, although their behavior hinted at the possibility of continuous laser action. As Bridges relates in his interview, Eugene Gordon demonstrated the first continuous-beam ion lasers at Bell Labs (Gordon and Labuda, 1964) by shrinking the discharge bore to a millimeter in diameter. This raised the current density to levels high enough to sustain continuous laser action.

The mercury-ion laser also led to development of metal-vapor lasers, both in neutral and singly ionized species. It inspired Grant Fowles, a professor at the University of Utah, to begin a systematic search for metal-vapor lasers with graduate student William Silfvast, who describes the results in his interview. Their first successes were in early 1965 with zinc and cadmium (Silfvast, Fowles, and Hopkins, 1966). They were concentrating on the metals, and initially they did not realize that the helium buffer gas they used to carry the discharge through the tube was essential for laser action on some transitions, including the 441.6-nm blue line used in commercial helium-cadmium lasers. Early metal-ion lasers operated only in pulsed mode. Silfvast made the first continuous helium-cadmium laser in late 1967 shortly after moving to Bell Labs.

While looking for a lead-ion laser at Utah, Silfvast and Fowles found an unexpectedly strong red line. They were surprised because their mirrors had very high losses at red wavelengths, which the laser had to overcome to operate. That red lead line was the first in a series of neutral-atom metal-vapor lasers. A group at TRG, which included Gordon Gould, picked up on the Utah work and demonstrated similar lasers in manganese and copper (Walter *et al.*, 1966). All these neutral metal-vapor lasers can operate only in pulsed mode because the lower laser level fills up, terminating the population inversion and the laser pulse in less than 100 nanoseconds. Atoms drop out of the lower laser level in about 100 microseconds, after which the laser can fire another pulse. Copper-vapor lasers are efficient producers of green light, so they have found commercial applications in recent years. Another member of the family, a gold-vapor laser emitting red light, also is offered commercially.

Chemical Lasers

All early gas lasers—helium-neon, rare-gas ion, carbon dioxide, and metal vapor—were powered by passage of an electric discharge through the gas. Discharge-pumping remains common for gas lasers, but alternatives exist. An attractive approach for high-power lasers is the use of a chemical reaction. That idea was first proposed by John C. Polanyi, a chemist at the University of Toronto, at a June 8, 1960, meeting of the Royal Society of Canada.

Polanyi is best known for his work on the physics of chemical reactions, which earned him the 1986 Nobel Prize in Chemistry. In 1958 he discovered that newly formed molecules emitted infrared light. Measuring this infrared radiation helped him understand how chemical reactions proceed. It also led him to predict that a chemical reaction could produce the population inversion needed for laser action. He made his proposal public in the interval between Maiman's demonstration of the first laser in May 1960 and its public disclosure in July. He submitted a paper outlining the concept to *Physical Review Letters*, only to suffer the same type of summary rejection as Maiman. His paper eventually appeared in the *Journal of Chemical Physics* (Polanyi, 1961).

Much research was devoted to chemical lasers during the next few years, and conferences on the topic were held in Moscow and San Diego. It was not until 1965, however, that J. V. V. Kasper and George C. Pimentel (Fig. 1.15) demonstrated the first chemical laser, a 3.7-micrometer hydrogen-chloride laser, at the University of California at Berkeley (Kaspar and Pimentel, 1965). Other demonstrations followed, eventually leading to continuous-output chemical lasers, which could be powered entirely from the energy produced in a chemical reaction without outside energy sources.

A few chemical lasers are used in research today. Their main attraction, however, is their potential to produce high power levels for use in laser weapons. Military research on high-power chemical lasers dates back to the 1970s. Chemical-laser battle stations are among the space-based weapon concepts that have been investigated as part of the Strategic Defense Initiative, or "Star Wars" program (Hecht, 1984; Strategic Defense Initiative Organization, 1989). The most powerful chemical lasers—MIRACL (Mid Infra-Red Advanced Chemical Laser) at the White Sands Missile Base

Figure 1.15 George C. Pimentel of the University of California at Berkeley, who made the first chemical laser (courtesy of the University of California, Berkeley).

and the Alpha laser built for the Star Wars program—can generate more than 1,000,000 watts of laser power in a continuous beam for a matter of seconds. Although that power sounds impressive, it still falls far short of the level needed for space weapons to defend against ballistic missile attack.

The Dye Laser

The use of organic dyes as the active media in lasers was first suggested in 1961. Dyes dissolved in liquids seemed attractive because they displayed strong fluorescence, but there was little effort to pursue the idea. Researchers may have been discouraged by unsuccessful experiments conducted by D. L. Stockman at General Electric. The time for dye lasers finally came in the mid-1960s, when Peter Sorokin at IBM demonstrated the first one, closely followed by independent groups at Hughes and the Max Planck Institute in West Germany.

Sorokin became interested in the spectral properties of organic dyes when he and John Lankard were testing them as Q switches to generate short, high-power pulses from ruby lasers. He first observed strong emission when illuminating chloro-aluminum phthalocyanine in ethyl alcohol with pulses from a ruby laser. When he and Lankard aligned mirrors with the dye cell, they saw infrared laser action in the dye so intense that it burned their photographic emulsion (Sorokin and Lankard, 1966). Sorokin's early-1966 results did not get wide attention quickly, and they were independently duplicated by Fritz P. Schaefer's group at the Max Planck Institute in West Germany (Schaefer *et al.*, 1966). Mary Spaeth and D. P. Bortfield (1966) at Hughes heard about Sorokin's work after setting up a similar dye laser experiment, and demonstrated laser action using a dye with quite different properties. Sorokin next tried pumping with a flashlamp rather than a ruby laser, and that approach also worked (Sorokin and Lankard, 1967).

The main value of the dye laser today is its ability to be tuned across a range of wavelengths, but the first dye lasers emitted a fixed wavelength, where the dye fluoresced most strongly. The first dye lasers emitted in the infrared, but by 1967 visible-wavelength laser action also had been demonstrated. That year also saw the first operation of a wavelength-tunable dye laser, by Bernard H. Soffer and B. B. McFarland (1967) at Korad. Like other lasers, the first dye lasers had mirrors at each end of the laser resonator, one of which transmitted some light as the laser beam. Soffer and McFarland replaced the totally reflecting mirror with a diffraction grating, which could be turned to different angles. Changing the angle of the grating changed the wavelength of light that the grating returned along the axis of the laser to the output mirror. This single wavelength was amplified in the laser, and appeared in the laser beam. This tunability has proved the most important property of dye lasers, and has led to their wide use in research laboratories. Schawlow used them extensively to study spectroscopy at Stanford University, work for which he received the 1981 Nobel Prize in Physics.

All early dye lasers were pulsed. It was not until 1970 that

Benjamin Snavely's group at Eastman Kodak operated the first continuous dye laser, which was pumped by a continuous argon-ion laser (Peterson, Tuccio, and Snavely, 1970). That type of dye laser has become important in many high-precision spectroscopy experiments. It also can be used to generate extremely short pulses, lasting as little as 6 femtoseconds (6×10^{-15} second).

High-Power Lasers for Weapons

The 1960s saw increases not just in the types of lasers available, but also in their output powers. Although some interest in high-power lasers came from potential industrial users, the big push came from the Department of Defense, which saw potential military applications. Some military developers envisioned weapons for use on the battlefield, but others had even greater hopes. In 1962 General Curtis LeMay suggested that lasers might defend against a nuclear missile attack (Stares, 1985).

Science-fiction writers had long envisioned beams of energy as weapons. The concept dates back at least to an obscure 1809 novel by Washington Irving, titled *A History of New York by Diedrich Knickerbocker* (Franklin, 1988, p. 21). The first well-known book to put such weapons in the hands of alien invaders was H. G. Wells's *The War of The Worlds*—written in the 1890s, and filmed in the 1950s. It's tempting to wonder if the movie version may have helped set the stage for military interest in laser weapons.

As soon as the laser was invented, the general press picked up the idea of laser weaponry, and produced some wildly speculative articles. One of them, titled "The Incredible Laser," appeared in a Sunday newspaper supplement late in 1962. The talk of plans for laser cannons and other science-fictional weapons was too much for Arthur Schawlow, and soon a copy of the article appeared taped to his door at Stanford, along with a message reading "For credible lasers see inside," shown in Fig. 1.16. Schawlow recalls, "Over the years, I gave away quite a few copies, and they turn up in all sorts of places. Even the author of the article eventually saw one of the posters, and wrote to say that he had not realized how appropriate the title had been" (Hecht, 1984, p. 26–27).

The
Incredible
Laser

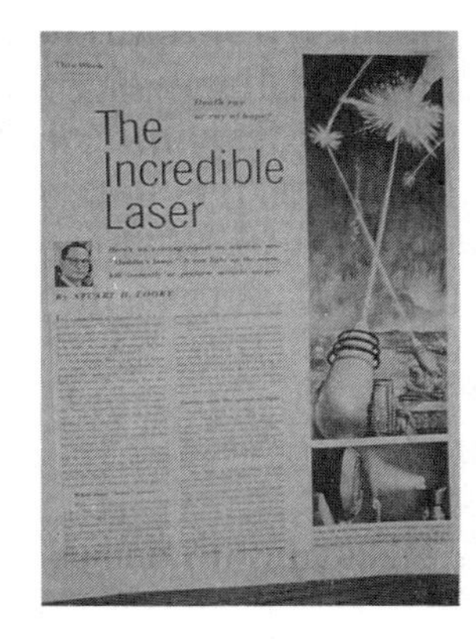

FOR CREDIBLE
LASERS SEE
INSIDE

Figure 1.16 Arthur Schawlow poked fun at wild speculations about laser weapons by posting this sign on his door at Stanford University in 1962 (courtesy of Arthur L. Schawlow).

Development of the carbon-dioxide laser raised hopes for powerful lasers, but early versions reached monstrous proportions. The laser shown in Fig. 1.17, built at Hughes Research Laboratories during 1967 to 1969 for the Defense Advanced Research Projects Agency, generated 1.5 kilowatts; it consisted of a small master oscillator, 16 meters of 25-millimeter amplifier tube, and an added 42 meters of 50-mm amplifier. A 8.8-kilowatt carbon-dioxide laser had a beam path 750 feet (230 meters) long. The huge size needed

Figure 1.17 Hughes Research Laboratories built this monstrous carbon-dioxide laser for the Defense Advanced Research Projects Agency in the late 1960s. The 10-watt master oscillator was followed by 12 meters of amplifier in a 25-millimeter tube and 42 meters of a 50-mm amplifier, delivering 1.5 kilowatts of average power. The tubes in this prototype, designed by William B. Bridges and Robert Hodge in 1967, were folded to fit on a 32- by 4-foot plywood optical bench. The prototype was finished and tested by Michael Smith; Howard Friedrich installed a cleaned-up model at the Rome Air Development Center in 1969 (courtesy of Hughes Research Laboratories).

to achieve high powers prompted one irreverent observer to say, "A laser big enough to inflict militarily significant damage wouldn't even have to work—just drop it on the enemy."

Military researchers soon began seeking new designs for carbon dioxide lasers, in hopes of reaching higher powers and efficiency. One method that has proved successful at modest powers is to apply short pulses of high voltage in a discharge perpendicular to the length of a laser tube. A. Jacques Beaulieu of the Canadian Defense Research Establishment at Valcartier began studying transverse discharges as a way to avoid the power limitations of early carbon-dioxide lasers, in which the discharge passed along the length of the tube. He began exciting the gas with a nominally continuous voltage, but arcing quickly stopped the laser action. Looking more carefully, he found that the laser was emitting most of its power as the voltage rose, so he began exciting the gas with shorter pulses, then varied gas pressure. In January 1968, he demonstrated the laser's ability to generate high peak powers in short pulses by destroying a detector (Potter, 1987). However, security and patent concerns delayed disclosure of the invention for two years (Beaulieu, 1970).

This type of laser works best when the gas mixture is at a pressure near one atmosphere, much higher than in standard carbon-dioxide lasers, which require low pressure to sustain a stable discharge through the gas. (The discharge length and pulse duration are short enough that stability is not a critical concern.) This type of laser is called TEA, for transversely excited, atmospheric pressure. The short, intense pulses have found some industrial applications, but did not reach powers high enough for weapon use.

A different breakthrough led to serious Pentagon efforts to build high-energy laser weapons. In 1967 researchers working on a classified military contract showed that expansion of a hot, high-pressure gas mixture containing carbon dioxide, nitrogen, and water could produce a population inversion. They were able to obtain laser high powers quickly from what they called a "gas-dynamic" laser. By the time the project was declassified in 1970, continuous power levels had reached 60 kilowatts (Gerry, 1970).

Ironically, Basov's group in the Soviet Union apparently was the first to propose producing population inversions by expansion of a hot gas (Basov and Oraevskii, 1963). Soviet researchers also built gas-dynamic lasers, which Prokhorov's group first reported in 1970 (Konyukhov *et al.*, 1970). The timing suggests that publication of the Soviet work may have led to declassification of the American research.

Soon all three armed services were testing powerful gas-dynamic carbon-dioxide lasers to evaluate the potential of laser weapons on the battlefield, in the air, or at sea. However, gas-dynamic lasers had their own limits, and the most powerful known to have been built was the 400-kilowatt laser installed in the Air Force Airborne Laser Laboratory, a military equivalent of a Boeing 707 (Hecht, 1984, p. 281). That system was built in the 1970s and flight-tested briefly in the early 1980s.

The highest continuous power claimed for any carbon-dioxide laser is 1 megawatt, from a laser at the Kurchatov Institute of Atomic Energy in Troisk, a scientific research center south of Moscow. That laser is powered by a novel type of electric discharge. Soviet officials showed it to visiting Americans in 1989 and told them it had been operating for several years (Hecht, 1989). Kurchatov officials told the visiting Americans that they had received military funding, and the laser may have been used to study the effects of high-power beams on potential targets. However, it lacked an external beam director for atmospheric tests, and was not "militarized."

Military researchers in the United States have generated even higher powers from chemical lasers. MIRACL, originally built for navy tests of laser defense of ships against missile attack, can generate 2.2 megawatts. The newer and more compact Alpha chemical laser, originally planned during the late 1970s and later built for the Strategic Defense Initiative, can generate comparable power, but the level is classified.

Other researchers have concentrated on generating extremely high peak powers in short pulses, lasting only about a nanosecond (10^{-9} second). Their main interest is in producing nuclear fusion by a process called "inertial confinement." The high-power laser pulse illuminates the surface of a tiny sphere filled with fusion fuel. The intense laser energy causes the target to implode, generating high temperatures and pressures for a very short period. The temperatures and pressures are high enough to allow nuclear fusion in the center of the target.

The concept of laser fusion originated in the 1960s. The Department of Energy (DOE) and its predecessors have conducted experiments for many years at the Lawrence Livermore National Laboratory, the Los Alamos National Laboratory, KMS Fusion Inc., the Naval Research Laboratory, and the University of Rochester's Laboratory for Laser Energetics. While the DOE professes an interest in civilian fusion applications, the program's official goal is to simulate nuclear explosions as part of the nuclear weapon program. The largest fusion laser in the United States, the neodymium-glass Nova laser at Livermore, also has been used in x-ray laser experiments, described in the interview with Dennis Matthews.

Excimer Lasers

By 1970, most lasers in commercial use today had been demonstrated, although important refinements remained to be made, especially for semiconductor lasers. One exception was the "excimer" laser, developed in the mid-1970s to fill an important gap in the laser spectrum by providing powerful pulsed sources of ultraviolet light.

The term "excimer" actually is a misnomer, coming from the origins of the device. Strictly speaking, an excimer is a complex of two identical atoms, which is stable only in an electronically excited state, such as a molecule containing two xenon atoms, Xe_2^* (the asterisk indicates the molecule is in an excited state). If the molecule releases its excitation energy, the two atoms are no longer bound together, and the molecule falls apart. A similar complex made up of two different atoms, such as xenon chloride, XeCl should be called an "exciplex."

The terminological borderlines became blurred in the laser world because excimer lasers that meet the narrow definition of the term were developed first. Lasers in which the molecules were made up of two different atoms, one a rare gas such as xenon, the other a halogen such as chlorine, came soon afterwards. They seemed similar enough to true excimer lasers that their developers appropriated the label "excimer," and it stuck. Molecules that exist only in the excited state are good laser candidates, because their lower laser level is a state that essentially does not exist—the molecules fall apart as soon as they lose their excitation energy. This makes a population inversion comparatively easy to produce, because the lower state is essentially always unpopulated.

The first proposal to use excimers as the active species in a laser dates from 1960, even before Maiman reported his laser, but it was not until 1970 that a group at the Lebedev Physics Institute—Basov, V. A. Danielychev, Yu. M. Popov, and D. D. Khodkevich (1970)—reported the first excimer laser, at 176 nm in the vacuum ultraviolet, from Xe_2 molecules. They produced the excited molecules by hitting liquid xenon with a beam of electrons. Two years later, another group reported the first molecular xenon laser at the same wavelength based on electron-beam excitation of xenon gas (Koehler *et al.*, 1972). High-power output from xenon was reported shortly thereafter (Hughes *et al.,* 1973), stimulating a round of intense interest that eventually wound down with the realization that the vacuum ultraviolet xenon laser would not be practical.

Interest in rare-gas halide lasers was triggered when Don Setser of Kansas State University reported emission near 300 nm from xenon fluoride in 1974. As excimer-laser pioneer J. J. Ewing relates in his interview, this quickly triggered a series of experiments with rare-gas halides. A number of laboratories mixed rare gases and halides, excited the mixtures with electron beams, and watched what happened. The initial results were observation of broad emission lines, mostly in the ultraviolet.

The first laser action was observed by Stuart Searles's group at the Naval Research Laboratory in Washington early the next year (Searles and Hart, 1975). Ironically, that first system was xenon bromide, which has found few uses because of its low efficiency. A couple of weeks later, Ewing and Charles Brau at the Avco Everett Research Laboratory in Massachusetts got xenon fluoride to lase at 354 nm (Brau and Ewing, 1975). Soon afterwards, they demonstrated xenon chloride and krypton fluoride lasers (Ewing and Brau, 1975). They got the strongest output from krypton fluoride at 249 nm, which is still considered the most efficient excimer laser.

All the early experiments were powered by firing a beam of electrons into the gas mixture. That approach has long been used to make high-power lasers, but it requires a large and expensive electron-beam generator. Fortunately, electron beams were not necessary. Later in 1975 Ralph Burnham and Nick Djeu at the Naval Research Laboratory modified a small commercial pulsed "TEA" carbon-dioxide laser and filled it with excimer gases. The device worked as an excimer laser, showing that discharge pumping was possible. That development led to present generation of commercial excimer lasers.

The Free-Electron Laser

The other important type of laser first demonstrated in the mid-1970s is the free-electron laser. Its origins can be traced back to a proposal by Hans Motz (1951), then at Stanford University, to produce millimeter waves by passing a beam of electrons through an array of magnets with alternating polarity. Such arrays are called "undulators" (or sometimes "wigglers") because their combined magnetic field varies along the array. Motz realized that the paths of electrons passing through this magnetic field would be bent back and forth, and that the bending would cause the electrons to emit radiation. (The same principle is behind synchrotron radiation, produced as the paths of electrons are bent by an accelerating magnetic field.) Motz demonstrated such emission in the millimeter-wave region, using a linear accelerator that was later used in small-scale free-electron laser experiments at Stanford.

The field lay largely dormant until John M. J. Madey became interested in the idea in the late 1960s. His first proposal for a free-electron laser was published in 1971 (Madey, 1971), and at about that time he began work at Stanford on demonstrating the concept, as he recalls in his interview. He concentrated on shorter wavelengths than had earlier researchers, who worked at millimeter or longer wavelengths. After several years, Madey's group demonstrated stimulated emission in a free-electron laser, by amplifying the beam from an external carbon-dioxide laser as it passed through an undulator magnet along with an electron beam (Elias *et al.*, 1976). They demonstrated the first free-electron laser oscillator the following year, obtaining peak power of 7 kW at 3.4 micrometers in the infrared (Deacon *et al.*, 1977).

Madey was not alone in the field, however. Independent of his work on infrared and visible free-electron lasers, a group at the

Naval Research Laboratory and Columbia University was working on similar concepts for generating millimeter and submillimeter wavelengths. Their contributions are sometimes missed because they did not use the label "free-electron laser," which Madey coined at Stanford.

There are important differences between the physics of long- and short-wavelength free-electron lasers, and they often are considered two different "regimes." In the short-wavelength regime, the dominant interactions are those of individual high-energy electrons in a low-current beam. Research on longer-wavelength devices has concentrated on the "collective" regime, in which individual electrons with low energy make up a beam with higher current, so the electrons are considered collectively. Following demonstrations of amplification of spontaneous emission at 400 micrometers and 1.5 millimeters, the NRL and Columbia groups collaborated on a collective-regime oscillator producing peak power of one megawatt at 400 micrometers (McDermott *et al.*, 1978).

An experimental lull followed those successes as theorists analyzed the free-electron laser concept. Much effort went into devising tests for claims that the free-electron laser would be able to generate high power output efficiently, and would be tunable in wavelength. The next round of experiments included operation of a visible-wavelength storage-ring free-electron laser at the University of Paris–South in Orsay; demonstration of kilowatt power levels at the Los Alamos National Laboratory; and various tests of physical predictions by other groups.

The Reagan administration's Strategic Defense Initiative poured well over $100 million a year into the development of high-power free-electron lasers in the middle to late 1980s. Their goal was to develop a high-power ground-based laser that could direct its output to orbiting mirrors, which could relay the beam to remote targets. SDI funded two separate development projects, one by the Los Alamos National Laboratory and the Boeing Corporation, the other by the Lawrence Livermore National Laboratory. The two programs used somewhat different technology, and SDI eventually picked the Los Alamos–Boeing approach for a larger-scale free-electron laser planned for construction at the White Sands Missile Range. However, cutbacks in the SDI budget and changes in the program's direction make it seem unlikely that the giant laser will actually be built in the near future.

Other researchers have turned to smaller-scale free-electron lasers for applications ranging from scientific research to medicine. Although free-electron lasers are expensive, they may be able to generate otherwise unattainable levels of power at some infrared wavelengths, valuable for certain applications. After lobbying by Madey, Congress set aside some "Star Wars" money for research into possible medical applications of free-electron lasers. SDI managers initially opposed the plan, but have since labeled it an example of the program's spinoffs.

The X-Ray Laser

Semiconductor and solid-state lasers advanced dramatically during the 1980s, but the only new type of laser to emerge was the x-ray laser. Its history is rather peculiar, even by laser standards. It includes, in addition to the saga of the laboratory x-ray laser at the Lawrence Livermore National Laboratory described by Dennis Matthews, a parallel program that remains highly classified and controversial.

Proposals for x-ray lasers appeared soon after those for visible lasers (Waynant and Elton, 1976), but actual demonstration of an x-ray laser proved exceptionally difficult. The fundamental problem is inherent in the physics. Population inversions on x-ray transitions have extremely short lifetimes. The only way to produce an x-ray population inversion is to concentrate a tremendous level of power on the laser medium. Different groups have found different solutions to the problem.

The first claim of success came in 1972 from the University of Utah (Kepros *et al.*, 1972) and only briefly approached the credibility level of cold fusion. John G. Kepros, then a junior researcher at Utah, found small, well-aligned spots on packets of x-ray film, and interpreted them as evidence of x-ray lasing from copper sulfate in Knox unflavored gelatin. However, no one was able to duplicate those results, and they were soon written off by the research community. Kepros ended up taking the blame for the incident, which most observers blame on over-optimistic interpretation of results rather than deliberate fraud (Hecht, 1984, pp. 123–124).

The Utah fiasco was not the only problem for x-ray laser research during the 1970s. Years of experiments had produced few encouraging results and much evidence that formidable difficulties awaited developers of x-ray lasers. The Defense Advanced Research Projects Agency, which had sponsored much research in the field, abandoned its program in late 1976 to shift funding to free-electron laser research. Some tantalizing hints of progress emerged from the Soviet Union in the late 1970s (Sobel'man, 1980; Hecht, 1984, p. 125), but American observers doubted they had demonstrated an x-ray laser. By about 1980, the topic virtually vanished from the open Soviet literature, either because they had given up or because it had become classified.

Ironically, only after support dried up did the first real signs of progress appeal. In 1980 Geoffrey Pert's group at the University of Hull in England reported laser gain at 18.2 nanometers in a highly ionized carbon plasma made by vaporizing thin carbon fibers with intense infrared laser pulses. Recombination of electrons with the carbon ions produced the 18.2-nm emission, which a detailed analysis concluded represented gain from stimulated emission (Jacoby *et al.*, 1981, 1982). The Hull group cautiously classed the wavelength in the extreme ultraviolet. However, the boundary between extreme ultraviolet and x-ray is hazy, and others did use the "x-ray" label because the wavelength was much shorter than previous lasers.

If the Hull results were encouraging, they were far from definitive. The calculated gain was low, and skeptics worried that it could have other causes. In addition, the experiments did not demonstrate oscillation, because cavity mirrors were not available for such short wavelengths.

The most dramatic and controversial x-ray laser story surfaced in early 1981. Quoting unnamed sources, *Aviation Week & Space Technology* magazine reported that physicists from the Lawrence Livermore National Laboratory had made a 1.4-nanometer x-ray laser by pumping the laser medium with x-rays produced by a nuclear explosion (Robinson, 1981). Although that article did not name the people involved, other sources indicated the concept was originated by Livermore physicists George Chapline and Lowell

Wood, and that the experiments were directed by Livermore physicist Thomas A. Weaver.

The *Aviation Week* story created a furor. Like virtually all aspects of the Department of Energy's nuclear testing program, the bomb-driven x-ray laser experiments were highly classified. Reporters trying to verify details were greeted by stony "No comment"s from Livermore, the Department of Energy, and the Department of Defense. *Aviation Week,* which normally follows up on stories it publishes, had nothing more to report, and refused to let *Omni* reprint an illustration from the article (Hecht, 1984, p. 130). Even today, government officials can do little more than confirm that bomb-driven x-ray lasers have been tested and that they produced the shortest wavelength on record, as Dennis Matthews does in his interview. No publications describing bomb-driven x-ray laser experiments have appeared in the scientific literature, although the scientists involved were profiled in a popular book, *Star Warriors*, by William Broad (1985).

The bomb-driven experiments showed that the pump power densities needed to produce an x-ray laser could be produced by sheer brute force. Matthews's group at Livermore took a more elegant approach to generating extremely high power densities. They focused short pulses from a massive laser built for fusion experiments at Livermore onto small areas of metal foils, producing plasmas and population inversions. (Pert's group also focused short laser pulses onto small targets to generate x-ray population inversions, but their laser was much less powerful, and their targets were fibers rather than foils.)

In October 1984, Livermore announced the success of Matthews's group in demonstrating what officials were careful to call the first "laboratory" x-ray laser. Matthews, Mordecai Rosen, theorist Peter Hagelstein, Weaver, and others used intense pulses from the Novette glass laser to generate stimulated emission from thin foils of selenium and yttrium. The selenium emission lines were at 20.6 and 20.9 nm; the yttrium lines were at 15.5 nm (Rosen *et al.*, 1985; Matthews *et al.*, 1985). The amplification was much higher than at Hull, and specialists consider the Livermore work the first definitive demonstration of x-ray laser action.

Matthews announced his results at a meeting of the Plasma

Physics Section of the American Physical Society in Boston. The Livermore scientists were surprised to find their Boston press conference joined by Szymon Suckewer of the Princeton Plasma Physics Laboratory, who reported gain on the 18.2-nm carbon line much stronger than that observed at Hull (Suckewer *et al.*, 1984, 1985). Like Pert, Suckewer used laser pulses to produce a carbon plasma, but at Princeton a magnetic field helped confine the plasma, making gain higher.

Developers of laboratory x-ray lasers have made continuing progress to shorter wavelengths and more efficient operation in the years since their first demonstration. Wavelengths have reached the 4-nm range, desirable for x-ray holography of living objects, as Matthews mentions.

The bomb-driven x-ray laser had become a center of controversy even before demonstration of the laboratory version. The nuclear experiments were originally funded from Livermore's discretionary funds, and some observers have interpreted the leak to *Aviation Week* as an effort to secure outside funding. Edward Teller, considered the father of the American hydrogen bomb, reportedly cited the x-ray laser as a dramatic breakthrough when he helped persuade Ronald Reagan, then president, to launch the program that became the Strategic Defense Initiative.

The bomb-driven laser quickly became one of the most controversial parts of "Star Wars." While Reagan was proclaiming that SDI was a "non-nuclear" program for defense against nuclear attack, critics noted that the bomb-driven x-ray laser was explicitly nuclear. Deploying it would require scrapping three major arms-control treaties.

Much more damaging are claims that bomb-driven x-ray laser advocates, particularly Wood and Teller, exaggerated the progress of the program. Some problems appear to date back to miscalibration of instruments used in early tests, so results appeared better than they should have. Later experiments showed that the bomb-driven x-ray laser did work, but not as well as had been thought (Smith, 1985a, 1985b).

Meanwhile, Wood and Teller became enmeshed in a controversy with Roy Woodruff, who as associate director for defense systems at Livermore was responsible for their program. Woodruff felt that statements Wood and Teller made to President Reagan and other high government officials were misleadingly optimistic about prospects for bomb-driven x-ray laser weapons. Teller definitely was optimistic; in a December 1984 letter, he told key Reagan advisor Paul H. Nitze that "a single x-ray laser module the size of an executive desk . . . could potentially shoot down the entire Soviet land-based missile force" (General Accounting Office, 1988). After complaints to Livermore director Roger Batzel went unheeded, Woodruff resigned as associate director and was demoted to a much lower position, essentially a form of internal exile (Morrison, 1990).

Woodruff filed a grievance about his treatment with the University of California, which administers Livermore for the Department of Energy, and sources within the university eventually leaked those documents, triggering investigations by Congress and the General Accounting Office. Ultimately, Woodruff's grievance was sustained, he was for the most part vindicated, and he was reassigned to a management position at Livermore. In May 1990, he left Livermore to manage the treaty verification program at the Los Alamos National Laboratory.

Meanwhile, the bomb-driven x-ray laser fell from favor for missile defense as careful technical evaluations revealed serious limitations. Funding for Department of Energy research for the Strategic Defense Initiative, dominated by bomb-driven x-ray laser studies, dropped from $360 million in fiscal 1987 to zero in 1991 (Morrison, 1991). In each of those years, Congress sharply cut administration requests, initially $603 million for 1977 and $192 million for 1991. However, the congressional cuts did not kill the program, because the Energy Department can fund continuing research from its general nuclear-weapons accounts.

The Laser Community

Laser technology has spread far beyond the small group of researchers who develop new lasers. During the past three decades, laser technology has found a tremendous range of applications. Lasers are used in scientific research ranging from ultraprecise measurements of time and energy levels to driving nuclear fusion reactions. In industry, lasers cut plastic and metal, drill microscopic holes in cigarette paper, and help align machine tools. In medicine, lasers treat blindness caused by diabetes, shatter kidney stones, perform surgery, and bleach away birthmarks. In offices, lasers print computer output and read and write data on optical disks. In stores, lasers read the striped symbols of the Universal Product Code. On farms and construction sites, lasers define surfaces that work crews use in grading land or mounting wallboard. Under the ocean, lasers relay signals through fiber-optic cables to link telecommunication networks on different continents. In homes, lasers play static-free music from compact discs.

A thriving industry has grown up to supply these many uses of lasers. As we saw earlier, the oldest companies date back to the early 1960s. Some companies have come and gone, and others are still being formed. Three laser pioneers interviewed in this book, Maiman, Gould, and Javan, have helped to start companies in the field. Sales of lasers per se, not including accessories such as external optics and parts handling equipment, run close to a billion dollars a year worldwide. The total value of products containing lasers, from welding systems to compact disc players, runs well into the billions.

Government and industrial laboratories around the world continue to carry out extensive laser research. AT&T, Bell Communications Research, and several Japanese companies have extensive programs to develop semiconductor lasers, optical communications, and optical computing. The U.S. Department of Energy has programs on laser-driven nuclear fusion, and laser separation of plutonium and uranium isotopes. The U.S. Department of Defense continues to study high-energy laser weapons as part of the Strategic Defense Initiative, although SDI emphasis is shifting to nearer-term technologies in the 1990s. The armed services have long used lasers to measure the ranges to potential targets, and to identify targets for smart bombs; laser-guided bombs were widely used in the Persian Gulf War of 1991. Lasers also are used in military training exercises, where soldiers zap each other with harmless laser beams, and sensors record "kills." NASA is working on laser systems for remote sensing from space. The list could go on for pages.

The Past Is Not Forgotten

A few laser pioneers have drifted away from the field. Townes's main activities now lie in astrophysics. Patel has become a research manager at Bell Labs. Before his retirement, Hall worked on problems relating to very-large-scale integration of semiconductor electronics. Gould, who retired at 65 in 1985 after he started receiving income from his patents, is involved with investment and high-technology management. Others remain deeply involved in laser development or applications. Madey, Matthews, and Silfvast still are working on laser development. Sorokin, Schawlow, Bloembergen, and Bridges are involved with laser applications.

The pioneers have received many awards for their work, only a very few of which are listed in the introductions to their interviews. Two of the professional societies in the field have named awards after pioneers. The Optical Society of America established the annual Charles Hard Townes award in 1980 for "outstanding experimental or theoretical work, discovery, or invention in the field of quantum electronics." Its recipients include James Gordon and Herbert Zeiger (who worked on the maser with Townes) in 1981, and Patel in 1982. The Laser Institute of America established the annual Arthur L. Schawlow medal for laser applications in 1982, and awarded the first medal to Schawlow "for distinguished contribution to laser applications in science and education."

To commemorate the twenty-fifth birthday of the laser in 1985, the American Physical Society, the Laser Institute of America, the Optical Society of America, the Lasers and Electro-Optics Society of the Institute of Electrical and Electronics Engineers, the IEEE Center for the History of Electrical Engineering, and the American Institute of Physics Center for the History of Physics launched the Laser History Project. It recorded a series of interviews with laser pioneers and compiled documents on laser history. The project also sponsored science historian Joan Bromberg to write a scholarly history of the laser, published in 1991 by MIT Press, titled *The Laser in America.*

Ironically, it may have been Gordon Gould's persistent efforts to secure a patent on the laser that helped heighten the laser field's sense of history. Gould's early claims were frustrated by a series of "interferences" with other patents. Although he continued to be-

lieve he had a right to patent coverage, he ran out of money and energy to fight the claims, and in the early 1970s enlisted the aid of the New York-based Refac Technology Development Corporation. Refac's resources were enough to do the job, and Gould was issued four U.S. patents:

- Patent 4,053,845, issued August 16, 1977, on "Optically Pumped Laser Amplifiers."
- Patent 4,161,436, issued July 17, 1979, on a broad range of laser applications.
- Patent 4,704,583, issued November 3, 1987, on electric discharge pumping of a laser.
- Patent 4,746,201, issued May 24, 1988, on the use of a Brewster angle window in a laser cavity (a technique that controls beam polarization and limits losses in the cavity).

Litigation began almost immediately after the first patent issued, and it continued for a decade. The Patlex Corporation, now based in Las Cruces, New Mexico, bought Refac's interest and part of Gould's in the patents, and prosecuted the infringement cases. The first case to come to trial was against the small General Photonics Corporation, which could not afford much defense in the face of the legal firepower brought in favor of the Gould patents. The Gould forces won that suit, but a series of delays kept a case against the Control Laser Corporation going until a 1987 trial, in which a jury upheld 8 of the 12 claims in Gould's optical pumping patent. Some companies had licensed the patents earlier, and that decision prompted the rest of the industry to go along. Gould now receives 20% of the patent licensing income, and Patlex receives 64%.

Laser Frontiers

After 30 years, the laser world still has frontiers, but they have evolved with the times. The pace of discovery of new types of lasers has slowed each decade. New lasers are still emerging from the laboratories, but they tend to be minor variations on existing themes, rather than radical new departures.

The most dramatic advances of the 1980s were in semiconductor lasers. Wavelengths were extended into the visible, emission bandwidths were reduced, output quality improved, and beam power greatly increased. Yet these changes have been made by fine-tuning structure and composition, not by finding a radically new principle of laser action. Quantum wells—thin layers with energy gaps between conduction and valence bands larger than those of the surrounding layers—may be the most important departure from previous semiconductor technology. They allow developers to tailor the characteristics of semiconductors to meet specific requirements by adjusting layer thickness and composition.

Optical-fiber amplifiers, which may change the rules for design of fiber-optic communication systems, likewise are soundly based in prior solid-state laser technology. They are simply glass fibers with cores doped with a rare earth element—typically erbium—which, except for the waveguiding provided by the fiber's inner core, function like solid-state lasers without mirrors to form a resonant cavity. Similarly, new solid-state lasers, such as alexandrite and titanium-doped sapphire, offer important new features, notably output that is tunable in wavelength. Yet they are in many ways variations on old themes.

That does not mean that innovation has changed, or that there is less room to explore in the world of lasers. It shows instead that researchers have mastered the basic principles of laser technology, and can draw on those principles to design, in a sense, lasers to meet system requirements. Consider, for example, the search for laser sources to pump erbium-doped optical amplifiers, which emit light at 1.55 micrometers. Researchers identified the two most promising pump bands as 0.98 and 1.48 micrometers, wavelengths where no good laser sources were readily available. Soon, however, developers had adapted quantum wells and other semiconductor laser technology to build sources for those wavelengths. Meanwhile, other researchers were devising ways to make erbium amplifiers that could be pumped at about 800 nm, in the range readily available from gallium arsenide semiconductor lasers. These developments promise users important new capabilities, but they do not open up whole new realms of laser physics.

Military laser research programs continue to press for higher laser powers, although budget cuts and changes in direction have stretched time scales for high-energy laser weapons for the Strategic Defense Initiative. SDI's Innovative Science and Technology Program is still looking for new lasers, but the program's major emphasis is on refining existing types so they can deliver higher powers. Meanwhile, energy researchers continue studies of laser drivers for initial confinement fusion, although, few observers expect laser fusion power plants before well into the twenty-first century.

Researchers in other areas of physics, chemistry, biology, and medicine are using lasers as sophisticated probes to monitor many processes. Developers have found ways to shorten laser pulses so they last as little as 6 femtoseconds—6×10^{-15} second—so they can observe ultrafast chemical and biological events. Laser beams can precisely measure frequency of atomic and molecular vibrations. Researchers have learned how to trap individual atoms with laser light, and cool them to temperatures close to absolute zero. They have found ways to generate "squeezed states," where energy or position can be measured more precisely than the laws of quantum mechanics might seem to indicate is possible. (Squeezed states do not violate the laws of quantum mechanics, but they do exploit a loophole.)

While we already know much about lasers, even more, remains to be learned. Like the explorers who mapped the world in the sixteenth and seventeenth centuries, the laser pioneers have determined the basic nature of the laser world. They have had the joy of being the first to see new vistas. Yet much more remains to be discovered by those who follow and can draw on the wealth of existing knowledge to study this new territory more closely than was possible at first glance.

References

Alferov, Zh. I., V. M. Andreev, D. Z. Garbuzov, Yu. V. Zhilayaeu, E. P. Morozov, E. L. Portnoi, and V. G. Trofim, *Fiz. Tekh. Poluprovodn. 4* 1826 (1970); translated in *Soviet Phys. Semiconductors 4* 1573 (1971).

Basov, N. G., and A. N. Oraevskii, *Soviet Physics-JETP 17* 1171 (1963).

Basov, N. G., and A. M. Prokhorov, "3-level gas oscillator," *Zh. Eksp. Teor. Fiz (JETP) 27* 431 (1954).

Basov, N. G., and A. M. Prokhorov, *Zh. Eksp. Teor. Fiz (JETP) 28* 249 (1955).

Basov, N. G., O. N. Kronkhin, and Yu. M. Popov, "Obtaining negative temperature states at the p-n junctions of degenerate semiconductors," *JETP 40* 1879–1880 (1961).

Basov, N. G., V. A. Danilychev, Yu. M. Popov, and D. D. Khodkevich, "Laser operating in the vacuum region of the spectrum by excitation of liquid xenon with an electron beam," *JETP Letters 12* 329 (1970).

Beaulieu, A. J., *Applied Physics Letters 16* 504 (1970).

Bell, W. E. "Visible laser transitions in Hg^{+}" *Applied Physics Letters 4* 34–35 (1964).

Benoit à la Guillaume, C., and Mme. Tric, "Les semi-conducteurs et leur utilisation possible dans les lasers," *Journal de Physique 22,* 843–836 (1961).

Bernard, Maurice G. A., and Georges Duraffourg, "Laser conditions in semiconductors," *Physics Status Solidi 1* 669 (1961).

Bertolotti, Mario, *Masers and Lasers: An Historical Approach* (Adam Hilger Ltd., Bristol, England, 1983).

Bloembergen, N., "Proposal for a new type solid state maser, *Physical Review 104* 324 (1956).

Brau, Charles A., and J. J. Ewing, "354-nm laser action on XeF," *Applied Physics Letters 27* 62 (1975).

Bridges, W. B., "Laser oscillation in singly ionized argon in the visible spectrum," *Applied Physics Letters 4* 128–130 (1964); erratum: *Applied Physics Letters 5* 39 (1964).

Broad, William, *Star Warriors* (Simon & Schuster, New York, 1985).

Collins, R. J., D. F. Nelson, A. L. Schawlow, W. Bond, C. G. B. Garrett, and W. Kaiser, "Coherence, narrowing, directionality, and relaxation oscillations in the light emission from ruby," *Physical Review Letters 5* 305 (1960).

Deacon, David A. G., L. R. Elias, J. M. J. Madey, G. J. Ramian, H. A. Schwettman, and T. I. Smith, "First operation of a free-electron laser," *Physical Review Letters 38* 892 (1977).

Dicke, R. H., "Coherence in spontaneous radiation processes," *Physical Review 93* 99 (1954).

Dicke, R. H, in P. Grivet and N. Bloembergen, ed., *The Coherence Brightened Laser in Quantum Electronics* (Dunod, Paris, 1964) p. 35.

Dicke, R. H., U. S. Patent 2,581,652, "Molecular amplification and generation systems and methods," issued Sept. 9, 1958.

Dupuis, Russell D., "An introduction to the development of the semiconductor laser," *IEEE Journal of Quantum Electronics QE-23* 651–657 (June 1987a).

Dupuis, Russell D., "Preface to 'Notes on the photon-disequilibrium amplification scheme (JvN), September 16, 1953,' " *IEEE Journal of Quantum Electronics QE-23* 651–657 (June 1987b).

Einstein, Albert, *Mitt. Phys. Ges., Zurich, 16* 18, p. 47 (1916); an English translation appears in B. L. van der Waerden, ed., *Sources of Quantum Mechanics* (North-Holland, Amsterdam, 1967).

Elias, L. R., W. M. Fairbank, J. M. J. Madey, H. A. Schwettman, and T. I. Smith, "Observation of stimulated emission of radiation by relativistic electrons in a spatially periodic transverse magnetic field," *Physical Review Letters 36* 717 (1976).

Ewing, J. J., and C. A. Brau, "Laser action on the $^2\Sigma^+_{1/2} \rightarrow {}^2\Sigma^+_{1/2}$ bands of KrF and XeCl," *Applied Physics Letters 27* 350 (1975).

Fox, A. G., and Tingye Li, "Resonant modes in a maser interferometer," *Bell System Technical Journal 40* 453 (1961).

Franken, P. A., A. E. Hill, C. W. Peters, and G. Weinreich, "Generation of optical harmonics," *Physical Review Letters 7* 118 (1961).

Franklin, H. Bruce, *War Stars: The Superweapon and the American Imagination* (Oxford University Press, New York and Oxford, 1988).

General Accounting Office, *Strategic Defense Initiative Program: Accuracy of Statements Concerning DoE's X-Ray Laser Research Program* (GAO/NSIAD-88-181BR, June 1988).

Gerry, Edward. T., "Gasdynamic lasers," *IEEE Spectrum 7* 11, p. 51 (Nov. 1970).

Geusic, J. E., H. M. Marcos, and L. G. Van Uitert, "Laser oscillations in Nd-doped yttrium aluminum, yttrium gallium, and gadolinium garnets," *Applied Physics Letters 4* 182 (1964).

Gordon, Eugene I., and E. F. Labuda, "Gas pumping in continuously operated ion lasers," *Bell System Technical Journal 43* 1827 (1964).

Gordon, J. P., H. J. Zeiger, and Charles H. Townes, "Molecular microwave oscillator and new hyperfine structure in the microwave spectrum of NH_3," *Physical Review 95* 282 (1954).

Hall, Robert N., G. E. Fenner, J. D. Kingsley, T. J. Soltys, and R. O. Carlson, "Coherent light emission from GaAs junctions," *Physical Review Letters 9* 366 (1962).

Hayashi, I., M. B. Panish, P. W. Foy and S. Sumelay, "Junction lasers which operate continuously at room temperature," *Applied Physics Letters 17* 3, 109–111 (Aug. 1970).

Hecht, Jeff, "House delegation sees Soviet 1-MW CO_2 laser," *Lasers & Optronics 8* 10, 19–20 (Oct. 1989).

Hecht, Jeff, *Beam Weapons: The Next Arms Race* (Plenum, New York, 1984).

Hellwarth, R. W., in J. R. Singer, ed., *Advances in Quantum Electronics* (Columbia University Press, New York, 1961) 334–341.
Holonyak, Nick, Jr., and S. F. Bevacqua, "Coherent (visible) light emission from Ga($As_{1-x}P_x$) junctions," *Applied Physics Letters 1* 4, 82–83 (1 Dec. 1962).
Hughes, William M., J. Shannon, A. Kolb, Earl Ault, and Mani Bhaumik, "High power ultraviolet laser from molecular xenon," *Applied Physics Letters 23* 385 (1973).
Jacoby, D., G. J. Pert, S. A. Ramsden, L. D. Shorrock, and G. J. Tallents, "Observation of gain in a possible extreme ultraviolet lasing system," *Optics Communications 37* 3, 193–196 (May 1981).
Jacoby, D., G. J. Pert, L. D. Shorrock, and G. J. Tallents, "Observation of gain in the extreme ultraviolet," *Journal of Physics B: Atomic and Molecular Physics 15* 3557–3580 (1982).
Javan, Ali, "Possibility of production of negative temperature in gas discharges," *Physical Review Letters 3* 87 (1959).
Javan, Ali, W. R. Bennett, Jr., and D. R. Herriott, "Population inversion and continuous optical maser oscillation in a gas discharge containing a He–Ne mixture," *Physical Review Letters 6* 106 (1961).
Johnson, L. F., and K. Nassau, "Infrared fluorescence and stimulated emission of Nd^{+3} in $CaWO_4$," *Proceedings IRE 49* 1704 (1961).
Kasonocky, W. F., R. Cornely, and I. J. Hegyi, "Multilayer GaAs injection laser," *IEEE Journal of Quantum Electronics QE-4* 4, 176–179 (1968).
Kasper, J. V. V., and G. C. Pimentel, "HCl chemical laser," *Physical Review Letters 14* 352 (1965).
Kepros, John G., Edward M. Eyring, and F. William Cagle, Jr., "Experimental evidence of an X-ray laser," *Proceedings of the National Academy of Sciences USA 69* 7, 1744–1745 (July 1972).
Keyes, R. J., and T. M. Quist, "Recombination radiation emitted by gallium arsenide," *Proceedings IRE 50* 1822 (1962).
Koehler, H. A., M. A. Ferderber, D. L. Redhead, and P. J. Ebert, "Stimulated VUV emission in high-pressure xenon excited by relativistic electron beams," *Applied Physics Letters 21* 198 (1972).
Konyukhov, V. K., I. V. Matrasov, A. M. Prokhorov, D. T. Shalunov, and N. N. Shirokov, "Gasdynamic CW laser using a mixture of carbon dioxide, nitrogen and water," *JETP Letters 12* 321 (1970).
Ladenburg, Rudolf, "Research on the anomalous dispersion of gases," *Physics Z. 48* 15–25 (1928).
Lamb, Willis E., Jr., and Robert C. Retherford, "Fine structure of hydrogen by a microwave method," *Physical Review 72* 241 (1947).
Lamb, Willis E., Jr., and R. C. Retherford, "Fine structure of the hydrogen atom, part I," *Physical Review 79*, 549 (1950).

Madey, John M. J., "Stimulated emission of bremsstrahlung in a periodic magnetic field," *Journal of Applied Physics 42* 1906 (1971).

Maiman, Theodore H., "Optical and Microwave-optical experiments in ruby," *Physical Review Letters 4*, 564 (1960b).

Maiman, Theodore H., "Stimulated optical emission in fluorescent solids, Part I, Theoretical considerations," *Physical Review 125*, 1145 (1961).

Maiman, Theodore H., "Stimulated optical radiation in ruby," *Nature 187* 493 (Aug. 6, 1960a).

Maiman, Theodore H., R. H. Hoskins, I. J. D'Haenens, C. K. Asawa, and V. Evtuhov, Part II, *Physical Review 125*, 1151 (1961).

Matthews, Dennis L., P. L. Hagelstein, M. D. Rosen, M. J. Eckart, N. M. Ceglio, A. U. Hazi, H. Mendecki, B. J. MacGowan, *et al.*, "Demonstration of a soft x-ray amplifier," *Physical Review Letters 54* 110–113 (1985).

McClung, R. J., and R. W. Hellwarth, "Giant optical pulsations from ruby," *Journal of Applied Physics 33* 828 (1962).

McDermott, D. B., T. C. Marshall, S. P. Schlesinger, R. K. Parker, and V. L. Granatstein, "High-power free-electron laser based on stimulated Raman backscattering," *Physical Review Letters 41* 1368 (1978).

Morrison, David, "The rise and stall of the X-ray laser," *Laser & Optronics 9* 8, 23–24 (Aug. 1990).

Morrison, David, "Congress becomes SDI's board of directors," *Lasers :& Optronics 10* 3, 21–22 (Feb. 1991).

Motz, Hans, "Applications of the radiation from fast electron beams," *Journal of Applied Physics 22* 527–535 (1951).

Nasledov, D. N., A. A. Rogachev, S. M. Ryvkin, and B. V. Tsarenkov, *Soviet Physics: Solid State* V4 782 (1962).

Nathan, M. I., W. P. Dumke, G. Burns, F. H. Dill, Jr., and G. Lasher, "Stimulated emission of radiation from GaAs p–n junction," *Applied Physics Letters 1* 62 (1962).

Patel, C. K. N., "Continuous-wave laser action on vibrational-rotational transitions of CO_2," *Physical Review A136* 1187 (1964).

Peterson, O. G., Sam A. Tuccio, and B. B. Snavely, "CW operation of an organic dye solution laser," *Applied Physics Letters 17* 245 (1970).

Polanyi, John C., "Proposal for an infrared maser dependent on vibrational excitation," *Journal of Chemical Physics 34* 347 (1961).

Potter, Roy, "An interview with A. Jacques Beualieu: The development of the TEA-CO_2 laser," *Optical Engineering Reports* pp. 1, 9A, 11A (Feb. 1987).

Quist, T. M., R. H. Rediker, R. J. Keyes, W. E. Krag, B. Lax, A. L. McWhorter, and H. J. Zeiger, "Semiconductor maser of GaAs," *Applied Physics Letters 1* 91 (1962).

Rabinowitz, Paul, S. Jacobs, and G. Gould, "Continuous optically pumped Cs laser," *Applied Optics 1* 511–516 (1962).

Ramsey, Norman F., "Experiments with separated oscillatory fields and hydrogen masers," *Science 248* 1612–1619 (29 June 1990).

Robinson, Clarence A., Jr., "Advance made on high-energy laser," *Aviation Week & Space Technology* 25–27 (23 Feb. 1991).

Rosen, M. D., *et al.*, *Physical Review Letters 54* 106 (1985).

Schaefer, Fritz P., Werner Schmidt, and Jurgen Volze, "Organic dye solution laser," *Applied Physics Letters 9*, 306 (1966).

Schawlow, Arthur L., and G. E. Devlin, "Simultaneous optical maser action in 2 ruby satellite lines," *Physical Review Letters 6* 96 (1961).

Schawlow, Arthur L., and Charles H. Townes, "Infrared and Optical Masers," *Physical Review 112* 1940 (1958).

Schawlow, Arthur L., and Charles H. Townes, U.S. Patent 2,929,922, "A medium in which a condition of population inversion exists," Mar. 22, 1960.

Searles, S. K., and G. A. Hart, "Stimulated emission at 281.8 nm from XeBr, *Applied Physics Letters 27* 243 (1975).

Silfvast, William T., G. R. Fowles, and B. D. Hopkins, "Laser action in singly ionized Ge, Sn, Pb, In, Cd, and Zn," *Applied Physics Letters 8* 318–319 (1966).

Smith, R. Jeffrey, "Experts cast doubts on X-ray laser," *Science 230* 646–650 (8 November 1985a).

Smith, R. Jeffrey, "Lab officials squabble over X-ray laser," *Science 230* 923 (22 November 1985b).

Snitzer, Elias, "Optical maser action of Nd^{+3} in barium crown glass," *Physical Review Letters 7* 444 (1961).

Sobel'man, I. I., "Atomic collision processes and UV and X-ray laser," in N. Oda and K. Takayanagi, ed., *Electronic and Atomic Collisions* (North-Hollard, Amsterdam, 1980) 75–80.

Soffer, B. H, and B. B. McFarland, "Stimulated emission observed from an organic dye—chloro-aluminum phthalocyanine," *Applied Physics Letters 10* 266 (1967).

Sorokin, P. P., and J. R. Lankard, *IBM Journal of Research and Development 10* 162 (1966).

Sorokin, Peter P., and M. J. Stevenson, "Stimulated infrared emission from trivalent uranium," *Physical Review Letters 5* 557 (1960).

Sorokin, Peter P., and M. J. Stevenson, in J. R. Singer, ed., *Advances in Quantum Electronics* (Columbia University Press, New York, 1961) 65; also, P. P. Sorokin and M. J. Stevenson, "Solid-state optical maser using divalent samarium in calcium fluoride," *IBM Journal of Research and Development 5* 56 (1961).

Sorokin, Peter P., and John R. Lankard, "Flashlamp excitation of organic dye lasers—a short communication," *IBM Journal of Research and Development, 11* 2, p. 148 (Mar. 1967).

Spaeth, Mary L., and D. P. Bortfield, "Stimulated emission from polymethine dyes," *Applied Physics Letters 9* 179 (1966).

Stares, Paul B., *Space Weapons and US Strategy: Origins & Development* (Croom Helm, London, 1985) 111.

Strategic Defense Initiative Organization, *1989 Report to Congress on the Strategic Defense Initiative* (SDIO, Mar. 13, 1989).

Suckewer, S., C. Keane, H. Milchberg, C. H. Skinner, and D. Voorhees, "Short review of recent soft X-ray laser development experiments at PPL," *Bulletin of American Physical Society* (Oct. 1984) 1211.

Suckewer, S., C. H. Skinner, H. Milchberg, C. Keane, and D. Voorhees, "Amplification of simulated soft x-ray emission in a confined plasma column," *Physical Review Letters 55* 1753–1756 (Oct. 21, 1985).

Townes, Charles H., "The early days of laser research," *Laser Focus 14* 8, 52–58 (Aug. 1978).

von Neumann, John, "Notes on the photon-disequilibrium amplification scheme (JvN), September 16, 1953" *IEEE Journal of Quantum Electronics QE-23* 659–671 (June 1987).

Walter, W. T., N. Solimene, M. Plitch, and G. Gould, "Efficient pulsed gas discharge lasers," *IEEE Journal of Quantum Electronics QE-2* 474–479 (1966).

Waynant, Ronald W., and Raymond C. Elton, "Review of short-wavelength laser research," *Proceedings of the IEEE 64* 7, 1059–1092 (July 1976).

Weber, J., "Amplification of microwave radiation by substances not in thermal equilibrium," *Transactions IRE Professional Group on Electron Devices PGED-3* 1 (June 1953).

White, A. D., and J. D. Rigden, "Continuous gas maser operation in the visible," *Proceedings IRE 50* 1697 (1962).

Wieder, Irwin, and Lynn R. Sarles, "Stimulated optical emission from exchange-coupled ions of Cr^{+++} in Al_2O_3" *Physical Review Letters 6* 95 (1961).

CHARLES H. TOWNES

Infrared and Optical Masers

Born July 28, 1915, in Greenville, South Carolina, Charles H. Townes received two bachelor's degrees from Furman University in 1935, one in physics and the other in modern languages. He earned his doctorate in physics from the California Institute of Technology in 1939. Townes spent the war years at Bell Telephone Laboratories working on radar systems, including one in the then-unexplored 24-gigahertz range. This interested him in microwave spectroscopy, a field he continued to study after joining the Columbia University physics faculty in 1948, where he remained through 1961, serving as department chairman from 1952 to 1955.

He conceived of the maser concept in 1951, and worked with postdoctoral fellow Herbert J. Zeiger and doctoral candidate James P. Gordon on experiments that led to demonstration of the first maser in late 1952. Later, he and Arthur L. Schawlow collaborated on developing the theory of "infrared and optical masers," or lasers. Townes was provost of the Massachusetts Institute of Technology from 1961 to 1966, and in 1967 was named university

Figure 2.1 Charles Townes with the second ammonia maser in spring 1955. This photo was on the cover of the June 1955 *Radio-Electronics* magazine (courtesy of C. H. Townes).

professor at the University of California, Berkeley. He has been an active advisor on many national issues.

Townes became professor emeritus in 1986, but remains active in research, primarily in astrophysics. His work includes microwave studies of molecules and molecular clouds, and infrared observations with high spectral and spatial resolution. Much of his work has been directed toward understanding the galactic center. He recently finished developing a pair of movable telescopes, which use spatial interferometry to achieve very high angular resolution of astronomical objects at infrared wavelengths.

In 1964 Townes shared the Nobel Prize in Physics with Nikolai G. Basov and Alexander M. Prokhorov of the Lebedev Physics Institute for "fundamental work in the field of quantum electronics which has led to the construction of oscillators and amplifiers based on the maser–laser principle." Other honors include the Comstock Award from the National Academy of Sciences in 1959, the Medal of Honor from the Institute of Electrical and Electronics Engineers in 1967, the C. E. K. Mees Award from the Optical Society of America in 1968, the NASA Distinguished Public Service Medal in 1969, foreign membership in the Royal Society of London, election to the National Inventor's Hall of Fame in 1976, the Neils Bohr International Gold Medal in 1979, and the National Medal of Science in 1982.

C. Breck Hitz conducted this interview on September 27, 1984, at the University of California, Berkeley. It was updated in June 1990.

Q: What were you doing before the maser was invented?

Townes: I was working on microwave spectroscopy, studying the interaction between microwaves and molecules. This involved looking in detail at the spectra and interpreting them in terms of molecular structure and nuclear moments. It was atomic and molecular physics that I was doing, using microwaves as the spectroscopic tool.

Q: It all developed from radar?

Townes: In a sense it did. During World War II, I worked on radar bombing systems, and became quite familiar with radar. In fact, one of the systems I worked on was at 1.25 centimeters, a system the Air Force wanted badly. But I believed this wavelength was likely to be absorbed by water vapor. I tried to persuade people that it wasn't going to work, but I was too young to be listened to and the decision had been made. Well, they went ahead and put it in the field, and it had no range because of water-vapor absorption. So all the equipment was discarded.

In the meantime, I thought a good deal about interactions between molecules and microwaves and realized that it provided the possibility of some very powerful spectroscopy. I thought that it might be possible to do things that nobody had been able to do before, using very high spectral resolution. We had the equipment, which was very cheap—it was sold in the streets of New York almost as junk! And so, in in a sense, the microwave spectroscopy work did derive from radar, and we used the surplus 1.25-cm equipment for a long time in microwave spectroscopy.

One centimeter was about as short a wavelength as could be reasonably used in the early days of microwave spectroscopy. But it was recognized that molecular interactions with microwaves become stronger at shorter wavelengths. So I was very eager to get down into the millimeter or even submillimeter range. The primary object of the work that led to the maser was to get shorter wavelengths so we could do better spectroscopy in a new spectral region and my goal was the far infrared or submillimeter region.

Q: Was it understood then that a population inversion was relatively easy to create?

Townes: Well, yes and no. At radio frequencies Pound and Purcell had inverted spin populations at Harvard, and that was known to physicists in the business. It was less well known to engineers.

But the effects that Pound and Purcell were observing were very weak, and nobody saw any practical application for them. I had thought about stimulated emission from time to time, as had other physicists, but had never done an experiment on it because it seemed too weak. I think it was not that people felt they couldn't get inverted populations. What wasn't realized was that the effects could be made big enough to give significant amplification. And nobody recognized the possible importance of oscillation and large amplification if that could be achieved. Nobody had thought of feedback. Putting a rather large population inversion in a resonant cavity with feedback is really what made the maser possible.

Q: Fabry-Perots had been around for a long time, of course.

Townes: Well, if you look back on it now, there's essentially nothing that wasn't known at that time by somebody. That's characteristic of many developments in science or technology. The breakthrough occurs when somebody realizes that a combination works in a way that no one had thought of before. Each individual element was known, and if someone had thought seriously in just the right direction, yes, they would have realized it could work. But nobody did.

So, while I understood that there was nothing particularly difficult about producing some population inversion, I didn't realize initially the potential it had as a radiation source. When I was at Bell Laboratories, I wrote a little essay for my bosses, trying to persuade them that microwave spectroscopy was of some importance. But I said that while molecules can produce radiation, the radiation would be weak, like blackbody radiation.

Now, that's good thermodynamics for thermal equilibrium. But then came the sudden realization that, wait—one doesn't have to use thermal equilibrium. If you don't use thermal equilibrium, then you can have inversion of population and (in principle) the radiation intensity can become enormous. Further, a resonant cavity could help achieve this. One aspect on which many good physicists, though not all, were confused was the coherence of stimulated radiation and hence the monochromaticity that such an oscillator could have.

Q: After the invention of the maser, did you immediately begin work on a visible maser—a laser? Or were there other projects that might have diverted you?

Townes: Well, as a matter of fact, at that point I took a sabbatical. I had just finished, with Art Schawlow, a book on microwave spectroscopy, and I'd also just finished a three-year term as chairman of the department of physics at Columbia University. I felt that microwave spectroscopy for physics was nearing completion. There was still a great deal of interesting chemistry that could be done with microwave spectroscopy, but the interesting things for physicists had been pretty well examined. It was time to think about what I wanted to do next.

So I cruised Europe and Asia—particularly France and Japan—for 15 months. I was looking at other fields, thinking about what I wanted to do and trying to decide whether to give up microwave spectroscopy. For a long time I'd been interested in radio astronomy, so I looked pretty hard at that field. But I was also thinking a lot about the maser. In France, one of my former students was working on a paramagnetic resonance with a long relaxation time. I realized that it could make a very good amplifier. We worked on that for a few months, the remainder of my time in Paris, and wrote a paper about it.

Then I left for Japan. Although I was still trying to decide what I should do next, the scientific work I did there was on the theory of amplification by stimulated emission. I worked with a couple of Japanese physicists, friends of mine there, in examining the discreteness of the resulting radiation field and the noise properties of maser amplifiers.

By the time I got home, I was convinced that maser development—later to be named quantum electronics—was the field I should continue with.

Q: Did you ever try to build a visible maser?

Townes: Yes, I did, at Columbia. After Art Schawlow and I wrote the original paper, which was published in 1958, I started to build a laser. We were looking at the alkalis, particularly cesium and potassium, essentially what we discussed in our paper. We planned pumping with other atoms, but basically we were trying to get an inverted population in one of the alkalis. One of my students and I, and later another associate, worked on it for about a year. I must have started in the fall of 1958, a few months before the paper came out.

But in the fall of 1959, I accepted a job in Washington. There weren't many scientists in Washington at that time. I felt it was important for scientists to tackle some of the problems like arms control, space work, and other technical issues that the country was struggling over at that time. That was shortly after Sputnik had gone up, you see. President Eisenhower had initiated the

President's Science Advisory Committee, and there was a great scramble to get scientists more involved in questions of national policy.

I was urged to go down to Washington—I felt a duty to go—so I did. I still continued to come up to Columbia on Saturdays, to take care of my students and continue some of my laser work. But of course it was a pretty small effort so far as I was personally concerned.

Q: Were you surprised when the Maiman laser first worked? Wasn't there a lot of doubt that a three level system like ruby could work at all?

Townes: I don't really think there was any doubt about a three level system in general. That's straightforward physics. Art Schawlow had also proposed a ruby system which was to work later, but Maiman's system was different and did involve levels which seemed more doubtful. I would have to say I didn't really expect making lasers to be as easy as it turned out to be. It seemed to me that people had done so much spectroscopy in the visible region, with gas discharges and other radiation sources, that somebody would have just stumbled into it if the laser were easy. So I felt it could not be easy, and everything must be planned carefully.

Art Schawlow and I based our specific calculations on the alkalis because the physics was well enough known that you could be sure a laser would work, if you did it right. That didn't mean it was the easiest case, but it was a case where we knew enough of the physics and enough of the parameters that it just had to work.

My style of physics has always been to think through a problem theoretically, analyze it, and then do an experiment which *has* to work. You analyze and duplicate the theoretical conditions in the laboratory until you beat the problem into submission, you see.

Now, there are other ways of doing physics; you can just try something and see whether it works. If it doesn't work, then go try something else. And that's a viable way of doing physics, too, but not the way I normally approach problems.

Q: Was Maiman's laser an example of this second kind of physics?

Townes: Well, maybe. Certainly all the parameters weren't well known. On the other hand, Maiman had been measuring properties of ruby. He had looked at some time constants and relaxation properties. But one couldn't have known for sure beforehand that it would work. Of course, it did, greatly to his credit. But as it turns out, almost everything works if you hit it hard enough.

Q: Has the alkali laser ever been "beaten into submission?" I mean, has the laser you and Schawlow originally proposed ever lased?

Townes: Yes, the cesium system has been made to lase. And I think most of the things we worked out in the 1958 paper surely can be done. There's just no longer great interest in the alkalis. There are so many better systems now, that just making another laser work isn't of great interest.

Q: What role did the Russians, Prokhorov and Basov, have in all this? Were the US and Russian efforts independent?

Townes: Basov and Prokhorov made an independent proposal. It's hard to know just when they learned of my own work. I first met Prokhorov at a meeting of the Faraday Society in Cambridge, England, in the spring of 1955. He was already familiar with our 1954 publication on the maser. But I knew nothing about their maser work. Prokhorov showed up at the meeting and talked about the possibility of making a maser. I was giving a paper on something else, but of course I was quick to comment on our working maser. We had a good chance to talk and compare notes. It was rare at that time to have a good occasion for meeting Russian scientists, and it was a privilege to have a chance to discuss things with Prokhorov. I knew some of his work in microwave spectroscopy from his published papers, but I had never met him.

I didn't know anything about Prokhorov and Basov's maser work prior to the 1955 meeting, although they had submitted a paper for publication in early 1954, and to what extent they knew about our work by then is hard to say. In addition to our 1954 publication, I had discussed our maser work in Japan in 1953,

Figure 2.2 Townes with the first ruby maser amplifier successfully used for radio astronomy in 1957 at the Naval Research Laboratory (courtesy of AT&T Bell Laboratories).

given a paper about the newly operating maser to the American Physical Society in the spring of '54, and we had written about the maser as early as the December, 1951 Columbia University progress report. Now, those reports were generally circulated only to a list of about 100 labs in the US and Europe, and normally didn't go to the Soviet Union. But it turned out they had always been put on the open shelves at least in the Harvard library. Lots of visitors had also been through our lab, so it's hard to know now how much got around.

But my belief is that the initial paper of Basov and Prokhorov on the maser was independent and disconnected from ours.

Q: Who are the other unsung heroes?

Townes: Well, I think Javan, Bennett, and Herriott were a very important group. They developed the helium-neon laser, which became the most common of all lasers [until millions of semiconductor lasers were made in the 1980s]. Their system involved quite different ideas from Maiman's, and if Maiman hadn't made the first laser, they would have shortly later. They may be somewhat overlooked for not being the very first.

Actually, they weren't second, either, although they're sometimes credited with that. They were third. There's another group that's also overlooked, Sorokin and Stevenson, of IBM. They made two lasers from rare-earth salts, somewhat along the lines of Maiman's, and I bet you haven't heard of that.

Q: I've always thought that HeNe was the second laser. A rare-earth salt? Was it optically pumped?

Townes: Yes; those were the second and the third lasers. They have had no substantial use. After that came the dark ruby lasers of Schawlow and others, and then Javan, Bennett, and Herriott with the HeNe laser. But I think this issue of exact time sequence is sometimes overemphasized. The role a contribution plays in the development of a field is perhaps more important.

There's another person whom you might consider unsung, a Russian scientist I have never met named Fabrikant. I don't think he's ever been out of the Soviet Union. He started in the 1940s, doing a thesis on inversion of population; as far as I know his ideas were original. He must have worked on the idea rather hard for a few years, though not much happened as a result. He applied for a patent in 1951, and when it was issued in 1959 it had a number of advanced ideas. But I just don't know enough about the Soviet patent systems to know what that means. I'm told that the Russian patent office allows quite a lot of changes between the time a patent application is filed and the time the patent is issued. Hence, what's published isn't necessarily what was initiated.

Nonetheless, it's pretty clear that Fabrikant initiated something in the field in 1951. That was pretty early, and I think he deserves some credit.

Q: Did Fabrikant interact with Prokhorov and Basov?

Townes: I don't know. Basov and Prokhorov, of course, deserve a good deal of credit, as do Maiman, Bloembergen, and Schawlow. But these people are generally well recognized for their contributions.

Q: What future do you see for laser isotope separation and inertial confinement fusion?

Townes: Well, lasers work very well in isotope separation. But I don't think that's going to change the world in a big way, it just makes the process cheaper. The only way laser isotope separation might really change the world is not particularly cheering; if it becomes so efficient that the production of militarily important isotopes is cheap and easy, then that's not so good. But it won't be all that cheap in any case; it'll just be cheaper than current methods.

As for fusion, yes, I'm convinced now that lasers represent one way of making fusion work. Whether that'll be the best way, I don't know. And whether, overall, the cost of building a plant that would produce fusion is going to be economical by comparison with other energy sources, I don't know. It's quite possible, but I think 20 years is the soonest that anything like that could be in use, and I believe it is likely to take longer than that.

Q: What about laser weapons and the strategic defense initiative?

Townes: I don't think anybody who has looked at it closely believes that optical lasers will solve the problem of defense against missiles. Maybe one can find somebody who believes they will, but the people I know don't believe it, nor do I.

X-ray lasers probably have some chance, but I don't really expect that they will work in the sense that they can overcome likely improvements in the offense. The trouble with the SDI program is not that we can't build fantastic things and shoot down almost any number of missiles, but rather that as one learns how to do that, the offense will learn too. And the offense seems likely to always have an advantage if offensive weapons are pursued as vigorously as defensive ones. So I don't think there's any way the defense can overwhelm the offense, unless further developments on the offensive side are neglected.

Q: Of course, that's not the way the world works.

Townes: That's not the way the world works, and we have to allow for that. For that reason I'm not very optimistic about the SDI program really providing us with any reasonably complete defensive protection.

Now that doesn't mean it's a useless program. You can stop some weapons, and if it's cheap enough to do that, then obviously that's useful, even though it doesn't solve the complete problem. There are intermediate possibilities that one can consider. But I am pessimistic about the possibility of any really complete defense.

Q: As you look back over the last 30 years, what about lasers has surprised you most?

Townes: Well there are several things that surprised me. One is how easy lasers really are to make. Considering the amount of work that had been done on optical spectroscopy before the 1940's, I'm still surprised somebody didn't make one accidentally.

Another thing that turned out to be better than I expected is the use of lasers to excite other lasers. By using lasers to excite lasers, you can do things which otherwise seem very difficult, like building tunable dye lasers, or making x-ray lasers. I think dye lasers are really spectacular, providing the remarkable amount of tuning one can get in the optical region. I had initially thought it would be very difficult. I thought you would have to be very inventive to do that. Fortunately, Peter Sorokin was.

Q: And what developments have pleased you most?

Townes: I'm very pleased, as well as surprised, by how useful the

Figure 2.3 Townes in his office at the University of California at Berkeley in February 1988 (courtesy of AT&T Bell Laboratories).

laser has turned out to be in the medical field. To me it's very rewarding to know that my friends' eyes have been saved by lasers. It's very different from the physics I do, more personal and emotional.

The laser is a natural for eye surgery, but it's also very successful in a number of other medical areas. That surprised me. I wrote a paper in the early days about possible medical and biological uses, and in that paper I was not optimistic about widespread medical applications. But doctors seem to be finding many more applications for it than I had supposed. They're investigating using lasers and fiber optics to remove plaque from the circulatory system and there are interesting ways of attacking cancer with lasers.

Communication by fiber optics is another area whose growth has been very pleasing to me, partly because that was something that was in our original patent. When Schawlow and I had a patent written up on the laser—we called it an "optical maser" then—we emphasized the possibility of communication partly because it was a Bell Labs patent. Now it's really coming into enormous use and making a very big difference in the cost of communication. Transoceanic fiber-optic cables seem to be cheaper than satellite communication. And, of course, they are much more secure.

I also am delighted and impressed by the large number of scientific contributions coming about as a result of quantum electronics and associated ideas. Very high precision, as well as attainment of extreme conditions such as very low temperatures and very short pulses, are some of the areas of beautiful work with lasers that it is a pleasure to recognize.

For some time I have argued that lasers really need not be expensive, and it is a delight to see the steadily decreasing costs of diode lasers as well as their increasing quality and flexibility.

Q: Have there been any laser applications or developments that have displeased you?

Townes: I guess one thing I don't like is the popular press harping on the laser as a death ray. Of course, death rays are fascinating to the human race; it's something they've speculated about long before the laser came along. Consider Jupiter's lightning bolt and Buck Rogers's ray gun—it's somehow fascinating to the public. But it's a distortion of the laser's overall potential for the newspapers to write so much about lasers as death rays, when in almost all cases they are very poor weapons. I guess I find any kind of distortion displeasing, but that one is becoming pretty boring. I wish the press would do better.

Q: What new directions do you expect laser technology to take in coming years?

Townes: Possible future directions for laser technology are still, as they were in the beginning, enormously varied. One need only recognize that lasers combine the field of light and optics with that of electronics. The importance and generality of each of these fields makes it obvious that quantum electronics, which combines the two, must have a very wide variety of applications. And so it does, and so it will. Perhaps the most obvious remaining need is for increased flexibility and quality and lower cost for lasers. I would like to see development of flexible lasers in the longer wavelength infrared and also at shorter wavelengths, including X-ray lasers. Both of these should be powerful scientific and technological tools.

Q: What are you doing now?

Townes: For the last 20 years, I have worked primarily in the field of astrophysics. While this may seem different, it is closely coupled with quantum electronics. For example, large numbers of naturally occurring masers have been discovered in astronomical objects and are playing an important role in the behavior of these objects and our understanding of them. Furthermore, much of the new instrumentation that I have been developing for improved astronomical measurements uses lasers in a variety of ways. This includes detection, measurement, precision controls, and optical components with unusual qualities. The work that I pursue most intensively at the moment is a mid-infrared Michelson-type interferometer, using two separate and movable large telescopes to achieve large baselines, giving very high angular resolution on astronomical objects. With them, we use laser heterodyne detection with CO_2 local oscillators and He-Ne lasers for precision control of all critical dimensions of the optical path. This is paying off and will, I hope, initiate a new and fruitful field of astrophysics.

An earlier version was published in *Lasers and Optronics* ®
(formerly Lasers and Applications) a Gordon Publications, Inc. publication.

ARTHUR L. SCHAWLOW

Origins of the Laser

After receiving a doctorate in physics from the University of Toronto in 1949, Arthur L. Schawlow spent two years as a postdoctoral researcher at Columbia University under Charles H. Townes before joining the technical staff at Bell Telephone Laboratories. He continued to collaborate with Townes while at Bell Labs, coauthoring a book on microwave spectroscopy and developing the principles of laser operation. In 1961 he left Bell Labs to become a professor of physics at Stanford University. Over the years his interests have evolved from developing new types of lasers to using lasers for high-resolution spectroscopy.

In 1981 Schawlow shared the Nobel Prize in Physics for his contributions to laser spectroscopy. He was elected one of six honorary members of the Optical Society of America, elected president of the Optical Society of America in 1975 and of the American Physical Society in 1981, named a fellow of the American Academy of Arts and Sciences, and elected to the National Academy of Science. The OSA awarded him the Frederick Ives Medal in 1976 "in recognition of his pioneering role in the invention of

Figure 3.1 Arthur L. Schawlow adjusts a ruby laser he made at Bell Labs in 1960 (courtesy of AT&T Bell Laboratories).

the laser, his continuing originality in the refinement of coherent optical sources, his productive vision in the application of optics to science and technology, his distinguished service to optics education and to the optics community, and his innovative contributions to the public understanding of optical sciences."

Schawlow received the third Marconi International Fellowship in 1977 and the Golden Plate Award from the American Academy of Achievement in 1983. In 1982 the Laser Institute of America presented him the first Arthur L. Schawlow medal for laser applications. In 1990 the American Physical Society established the Arthur L. Schawlow prize for laser physics. In 1991, Schawlow was elected an honorary member of the Royal Irish Academy.

C. Breck Hitz conducted this interview in early 1985 at Stanford University. It was updated in July 1990.

Q: Did your interest in science and optics begin early in life?

Schawlow: I would say so. As a boy, I was always interested in technical things, particularly radio, which was the term at that time for what we would now call electronics. It's hard for a person nowadays to imagine the excitement about radio in the 1920s, when I was a little boy, and into the early '30s. When I was about five years old, we got our first radio set, which had a big old horn loudspeaker. It needed an armful of batteries to run it. I remember particularly one year just before Christmas, when the local department store had a series of broadcasts of Santa Claus' progress from the North Pole to their store. All the kids on the block would come around to listen to this on our new radio.

I remember that at least once a week the newspapers used to have a column, a technical column, on radio: How to build your own one- or two-tube set, or crystal set, that sort of thing. It was something like the excitement over microcomputers during the last few years.

Q: That influenced you in your career choice?

Schawlow: Well, I was interested in that and also in mechanical things. I had a Meccano set, and I used to go to the library and bring home several books and read them. That's all there was to do in the summer. I went through nearly everything on the engineering and technical side of the public library.

I wanted to be a radio engineer, but there were two problems. When I finished high school, I was only sixteen years old, and you couldn't get into engineering unless you were seventeen. And there was a more serious problem. This was 1937 and the middle of the depression. I didn't have the money to go to university unless I got a scholarship. But there weren't any scholarships in engineering. I didn't quite know what to do, whether to take another year and try to earn some money, or what. Just for practice, I took the examinations for scholarships and was somewhat surprised when I got one in mathematics and physics.

I decided, "Well, that's near enough to engineering." So I went into mathematics and physics.

What I really expected to do, as practically everybody in our class did, was to teach high school physics. We had an honors course system at the University of Toronto, whereby you chose a major right away; the course was pretty narrowly focused from there on. Your first year was practically all mathematics and physics, and then later you'd branch out into specializing in mathematics or applied mathematics or astronomy or chemistry or physics. I moved into a physics specialty.

I think all of us in that class had high school teaching as a goal. But by the time we graduated Canada was at war, and none of us actually went into high school teaching.

Q: People wound up in the armed services instead?

Schawlow: That's right. Some of them served in the war as radar officers, or worked in war factories, or, as I did, taught courses for the armed services at the university during the war. Then I worked for a year on microwave antennas at a radar factory.

Q: Were you planning on graduate school then?

Schawlow: You know, that wasn't automatic the way it is for a lot of people nowadays. My family didn't have a lot of money. My father was an insurance agent, and the people that lived around us were clerks and bus drivers, that sort of thing. Not laborers, but still fairly ordinary working people. Most of them had never gone to university, much less graduate school. In fact, very few people in my high school class went to university, maybe two or three out of fifty or so.

So the thought of going on for graduate work and a PhD or research career was just really too much. I hoped I could get into some radio work, somehow, but I didn't really see it very clearly. The fall-back position would be high school teaching.

Q: How much time elapsed, then, between the time you graduated from college and the time you started graduate school?

Schawlow: I graduated in 1941, in February. Then I started back to graduate school in 1945. But I did get a masters degree earlier. I did a little research and got a masters degree in 1942. Then I didn't get PhD until 1949.

Q: Then what did you do, after you finished your PhD?

Schawlow: I wanted to do basic physics by that time. About a year or so before I got my PhD, there was a meeting in Ottawa of an organization called the Canadian Association of Physicists. I.I. Rabi from Columbia University gave an invited talk, and it was marvelous because he described the wonderful discoveries that Willis Lamb and Polykarp Kusch had made recently, for which they later got Nobel Prizes. I really wanted to go to Columbia more than anything else in the world. So I wrote to Rabi and he wrote back suggesting that I apply for the Carbide and Carbon Chemicals Corporation fellowship for the application of microwave spectroscopy to organic chemistry.

Well, frankly, my interest in organic chemistry was zero. I knew nothing about it. I'd never even had a course in it. But I was interested in microwaves. Actually, we'd had a klystron at the student lab in Toronto, and I'd had a chance to play with it. Besides, I had done microwave work for a year during the war.

The fellowship involved working with some guy whom I'd never heard of named Charles Townes. But I wanted to go to Columbia, so I applied for it anyway. And I won it.

Well, working with Townes at Columbia was a marvelous thing. There were no less than eleven future Nobel Prize winners at Columbia then, including Hideki Yukawa, who got his Nobel a few months after I arrived. Aage Bohr, that's Nils Bohr's son, was there, as were Townes, Polykarp Kusch, Willis Lamb, James Rainwater, Val Fitch, Leon Lederman, Melvin Schwartz, and Jack Steinberger. It was a very exciting compared to Toronto, which was somewhat of a backwater then. Toronto's much better now, but it had been badly damaged by the depression and hadn't really restored itself during the first few years after the war.

Q: How long did you stay at Columbia?

Schawlow: I left in 1951, after two years there. I started a book on microwave spectroscopy with Charlie Townes, but we didn't really finish it until 1954. Even after I'd left Columbia, I'd drive in nearly every Saturday and work on the book.

Q: What else were you doing in the meantime?

Schawlow: Well, I was getting married and I needed to have a job. Although there had been lots of academic jobs in 1949, there weren't so many in 1951. My wife wanted to stay near New York where she was studying singing.

So when I was offered a job at Bell Labs, it seemed like a good arrangement. John Bardeen had gotten interested in superconductivity and he wanted someone to help with his experiments. Well, I had absolutely no background in low temperatures or superconductivity, but there weren't too many PhDs around then and they offered me the position.

Unfortunately, by the time I got there Bardeen had decided to go to the University of Illinois. So there I was all alone, trying to do something on superconductivity. I had to learn all about solidstate and superconductivity by myself. But I taught solidstate physics courses for their incoming engineers, and that gave me a chance to learn some of the stuff I should have learned before.

Q: Were you involved in any of the early maser work going on about that time?

Schawlow: No, I didn't work on masers at all. I kept an eye on it, though, and I knew it was going on. In fact I had witnessed Charlie's notebook before I left Columbia. He told me about the maser idea, and it sounded good to me. If I had stayed at Columbia, I probably would have worked on it with him. But it was his idea.

By the summer of 1957 I was beginning to wonder seriously if there was some way you could get much shorter wavelengths out of some kind of maser. I wasn't really thinking about visible light yet, but of far infrared.

Charlie was consulting with Bell Labs by then. During lunch one day in early autumn we got to talking seriously about making some kind of an infrared or visible maser. We decided to put our heads together and see whether we could identify the problems and solve them.

He had one good idea right away. He thought that, instead of trying for the far infrared. we should go to the near infrared where we knew a lot more about the properties of materials. Spectra in the far infrared were pretty much uncharted at that time, you know. He told me about a scheme he'd worked out with thallium,

Figure 3.2 Schawlow demonstrates an early ruby laser with a linear flashlamp and cylindrical reflective cavity at Stanford University in 1962 (courtesy of A. L. Schawlow).

but I looked at it and decided it wasn't going to work. If I'm remembering right, the problem was that the bottom level would empty more slowly than the upper one would fill.

Well, we realized the first problem was to get enough excited atoms or molecules. In the optical region, we faced the problem that the lifetime of an excited state is limited by spontaneous emission. But Charlie had the maser equation, and we just kind of worked it around various ways. We saw that the lifetime, the oscillator strength, didn't really matter. If the transition had a low oscillator strength, the excited atom would last longer but each one would give less gain. With a high oscillator strength, you get more gain per atom but the rate at which you supply them would be just the same. But the branching ratio did matter, that is, the fraction that would radiate the desired wavelength.

We looked up oscillator strengths in various tables, and of course there was a lot of data about the alkalis. I finally picked potassium for a stupid reason: I had a visible-wavelength Hilger spectrometer which I bought when I started working on superconductivity, to measure the thickness of thin films by multiple-beam interferometry. I could do things in the visible; it was the only optical equipment I had. The first and second line of the potassium spectrum happened to be in the visible. It's the only one of the alkalis that had that. If I had known how rotten potassium is in some other ways—it's very reactive and easily quenched—I'd probably have picked something else. But we worked out a calculation showing that we could get enough excited atoms in potassium.

The other question was: what about mode selection? We never seriously considered trying to make a resonator as small as a wavelength. You wouldn't have room to put any atoms in the thing. We had in mind some kind of a large, oversized resonator, but Charlie didn't think we had to worry too much about what the modes would be like. He said, "Well, probably some mode or modes will have higher Q than others and they'll sort of pick themselves out and dominate. Or maybe it'll be jumping around rapidly, among a relatively small number of modes."

But I wasn't entirely satisfied with that, and Martin Peter, a physicist at Bell Labs, kept telling us that we ought to find a way of selecting one mode. I had all sorts of schemes to do that. There is a page in my notebook here, it's dated January 28, 1958.

You know, I didn't put many things in my notebook. I really was very poor at that. I was at Bell Labs for ten years, and in that time I went through only one and a half notebooks. It seems like every time I wrote something down in my notebook it was wrong, and only the things I wrote down on scraps of paper were worth anything!

Anyhow, on this particular page I had written something about mode selection. I was thinking about using a diffraction grating for walls of the resonator, to give some selectivity. Of course, people did that ten years later in tunable lasers. I had a lot of schemes of that sort.

Eventually I got the idea that all we had to do was select a direction, and to do that we should throw away most of the resonator and just leave a little piece at each end. I thought, "Well, at least it will define the direction to within the angle subtended at one mirror by the other." When I mentioned it to Charlie, he said, "It'll be better than that because the waves go back and forth a number of times and you'll get higher selectivity." It seems very obvious when you think of it that way now—but many things are very obvious after they work.

Well, our colleagues who knew about microwave resonators didn't believe that. They wanted a calculation of some kind, which I wasn't able to do. They thought there might be some modes where the electric field was along the axis of the resonator, as there are in microwaves. Well, we couldn't answer that fully, but Charlie did some calculations with a diffraction model and showed that those modes probably wouldn't oscillate. Of course, the full theory was worked out a year or so later by Fox and Li.

You know, it sometimes annoys me a little bit when people say I rediscovered the Fabry-Perot resonator. Nothing of the sort—I'd known Fabry-Perot resonators all along because I did my thesis with one. But this was something different, because diffraction has nothing to do with a Fabry-Perot, and here it was the diffraction that gives you the mode selection. And it was really, as far as I know, the first open resonator specifically intended to select a mode.

At that point, we thought it best to publish the idea and let other people go ahead and try to build a laser. I didn't even try to build one at first. I kept on working on superconductivity, although after a little while I went to the boss and said, "I think maybe it would be good for me to stop working on superconductivity and work on optical properties of solids."

Q: You were still at Bell Labs?

Schawlow: That's right. It was a wonderful place. I just had to say I thought I should stop what I was doing and start working on optical properties of solids, with a view toward developing a laser, and that was all it took.

The term "laser," incidentally, is often misattributed. The name "maser"—an acronym for microwave amplification of stimulated emission of radiation—was invented after the first device was built. Then when masers were developed in other spectral regions, people just naturally started tossing around all sorts of variations. I've seen in print "iraser," for an infrared device, "raser" for a radio-frequency device, and so forth. Laser was just another obvious variation. But it was the one that got popularized. Eventually we all shifted over to laser. But the important thing to realize is that it was a variation on Townes' original maser.

Q: In his interview, Gordon Gould told us he had invented the term.

Schawlow: He was one of the first to push that name, maybe. I think I heard him use it as early as '59, but I don't think he was the first to use it. He may have forgotten how the name really came about.

Q: While we're talking about people involved with early laser work, are there people who made significant contributions to your early work who aren't particularly remembered?

Schawlow: A lot of people were involved in the early work, and I don't think I can name them all just off the top of my head. Ali Javan thought of the idea of the gas discharge laser, and that's certainly turned out to be a pretty good idea. Bill Boyle at Bell mentioned that you could use a gas discharge but it would be better to use a semiconductor, which is a solid-state analog of a gas discharge. We didn't mention that in our paper because it was Boyle's idea, not ours. Unfortunately, he never published it either, although he got some kind of a patent on a solid-state laser.

But Javan, together with Bennett and Herriott, got helium-neon working. I think the gas laser owes nothing at all to Maiman's work, although I certainly don't want to run down his contribution, which was very important. But there still would have been lasers if Maiman had never lived, because Javan's HeNe laser was developed quite independently. Both of them did use our resonator structure, though: two little mirrors facing each other.

Q: Maiman's laser—the first laser—operated on the R-lines of chromium-doped ruby. You had predicted that those lines weren't suitable for laser action, hadn't you?

Schawlow: Yes, it's in print. I said that in 1959. I outsmarted myself because I wasn't being quantitative.

I had the feeling that, since lasers had never been made, it must be terribly difficult to make one. You had to give yourself every advantage. In particular, it always seemed to me that you'd have to have a four-level system. With a three-level system like ruby, you'd have to lift half of the atoms out of the ground level before you got any gain at all.

But I still had ruby in mind for a laser. This was at Bell Labs, and the feeling was "Anything you can do in a gas, you can do better in a solid." So I suggested a scheme for how ruby could work. We had found out that the ground level could be split by the exchange interaction between pairs of chromium atoms if the chromium concentration were high enough. And the splitting could be as large as several hundred wavenumbers. That meant that we could operate at very cold temperatures to depopulate the higher level, and in essence have a four-level system.

I tried making such a laser in a very half-hearted way. I used a dark ruby rod—you want a darker ruby because it has more of the pairs—and a little 25-joule flashlamp. Nothing much happened, so I put it aside for the time being.

That was the reason for my off-the-cuff remark, about the R-line being unsuitable for laser action. I thought I was being clever, but, as I said, I outsmarted myself. But I guess I was the first to propose ruby. It may well be that I'd drawn Maiman's attention to ruby by mentioning, in various places, that we might be able to use this dark ruby. I understand that—unfortunately—my prediction caused Maiman some difficulty. He complained that people took it more seriously than I intended it to be.

But I didn't know Maiman at all well. I'm not even sure if I had met him or not at that time. So I certainly never said that his idea wouldn't work, because I'd never heard of his idea.

Q: Did you ever try your dark ruby laser again?

Schawlow: In 1960, after Maiman's laser had worked, I again went to the boss and said, "You know, I really believe I could make that dark ruby laser work. Should I take the time to try it?" And he said, "You owe it to yourself." So we did, and very quickly it did work. And we published it . . . submitted it in late 1960.

Q: Thirty years have gone past since the events we've been a discussing. What laser applications during those years have seemed most exciting to you?

Schawlow: Well, one gets certain satisfaction from the medical applications of lasers. They're always nice to point out. One of the very first laser applications, within a year or so of Maiman's laser, was repairing detached retinas. Neither Charlie nor I had ever heard of a detached retina back in 1960. If we had been trying to develop new medical techniques, we wouldn't have been fooling around with stimulated emission from excited atoms. But the eye doctors were already using xenon arc flashlamps to produce scar tissue on the retina to prevent detachment. It was obvious to them that once you had a brighter light, it was worth trying. So they did and it worked.

In the early days I used to get frantic calls and letters from people wanting me to cure them, or their relatives, of cancer. Of course, I'm not a medical man anyway. I always tried to be careful to say that the laser wasn't a cure for cancer. I often joked that, no matter what you told the press about lasers, it always came out as a "death ray" or a cure for cancer—or both! I even had somebody write and want me to cure their dog's cancer. I passed that one on to Leon Goldman, and actually he gave a very reasonable answer to what could be done for a dog's tumor.

You know, people sometimes ask me if I'm ashamed of inventing a death ray. I don't believe we have laser death rays . . . although I'm not sure it would be bad if we did. But we don't, at least not to

Figure 3.3 Schawlow in his Stanford Laboratory in 1988 (courtesy of A. L. Schawlow).

do the job that people would like done, for missile defense and things like that.

Q: There's a lot of talk in the laser community these days about the Strategic Defense Initiative. How realistic is SDI? Is that a reasonable thing to be doing with lasers?

Schawlow: Well, I don't know. I don't do military work, and I haven't since World War II. But I do think a good defense would be very worthwhile. It would be good if they could do something to end this balance of terror, which I think is inherently unstable, especially if more countries get atomic weapons. Some of the smaller countries certainly act in a very irresponsible way. At least we and the Russians can hold each other responsible. You know who it is you're dealing with. But if we suddenly had a loose missile come in, from God knows where, some kind of defense would be desirable.

I'm on the American Physical Society's directed energy weapon study group review committee. But my only function there is to make sure that the report makes sense to a person like myself, who is not involved in the SDI program.

I think, from all I've read, though, that no matter how much money you put in now, you couldn't do it. It's questionable whether you're ever going to be able to do it satisfactorily. Some people argue that the offense can always overwhelm the defense with more cheap missiles, and I don't know that they're wrong.

Q: What other laser applications have you found gratifying?

Schawlow: There have been a lot of scientific applications. This was one of the few things we had in mind back at the beginning. When we went into this, I understood the history of radio and how people had gone from broadcast band to shortwaves, and found they could talk around the world with them. And then very high frequencies made possible broadband broadcasting, like television and FM. Then, at even higher frequencies, microwaves were no good for broadcasting but great for radar and point-to-point communication. I was certain that, if you could make coherent radiation at shorter wavelengths, there would be uses for it.

We had a vague idea about laser chemistry. We also thought you could probably do some spectroscopy with lasers, if you could learn to tune them. But I didn't think it would be easy to tune very far. Well, the first few years you couldn't tune them very far, but in the late 1960s Peter Sorokin and J.R. Lankard—and, independently, Fritz Schaefer and his colleagues—developed the dye laser. That meant we could do some laser spectroscopy, and it really opened up a whole new world for us. I found that very exciting. It gave us new levels of sensitivity and resolution. It meant you could detect single atoms and photograph single atoms, a marvelous thing.

Q: I guess the only place where laser photochemistry is really practical so far is in nuclear fuel processing.

Schawlow: It may be practical there. I think they're still debating whether that's the most practical way to enrich uranium. [The US Secretary of Energy decided in favor of the laser isotope separation process on June 5, 1985—*Ed.*] I stopped working on photochemistry because I was afraid I might find something that would make it too easy. Obviously, the big countries need isotopes separated for atomic power and also for weaponry. But it just scares me, again, if irresponsible groups, terrorists or some irresponsible countries should get access to separated and purified uranium. So far, it's difficult to enrich uranium, and I hope it stays difficult. But I'm not going to look into it.

Q: With all the work that's gone into laser isotope separation, you'd think that somebody would have found an easy way if there were one.

Schawlow: You would have thought that about lasers, too. It really was surprising how easy it was to make them, and also how quickly they became powerful.

Q: What other things about lasers have surprised or impressed you during the past 30 years?

Schawlow: Well, how easy they were was certainly one surprise. We were pretty confident of what we'd written, but you're never sure that you haven't overlooked something until it actually works. So I guess I was a little surprised that it turned out to be as easy as we'd said it would be.

It also surprised me that the first one was so powerful: 1000 watts. With our background in communications and microwave spectroscopy, we thought, "Well, if you can do it at all, maybe you'll get a microwatt or something like that."

Very many clever people have been attracted to this field, and they have produced ideas, discoveries, and inventions that could not have been foreseen. The introduction of tunable dye lasers had a great effect on my own research. There may well be a similar impact from the advances in visible semiconductor diode lasers. The advances in ultra-high-power lasers for fusion research and ultra-short-pulse lasers are also impressive. Optical solitons and their application to long distance communications were quite surprising.

Q: What has disappointed you about lasers?

Schawlow: Some things do frustrate me. For instance, here we are, 30 years after the first laser, and we still don't have an efficient, high-power, visible-wavelength laser. It would be so much better for metalworking or for chemistry than infrared lasers. We just don't have anything that can give us more than 20 watts or so in the visible. And those lasers are so inefficient that practically all the power goes into the device, and it soon burns itself up. We spent a lot of money in this country replacing argon laser tubes.

Q: What new directions do you expect laser technology to take in the coming years?

Schawlow: The easiest way, and perhaps the only reliable way, to answer is to project trends already under way. Thus we can expect to see progress in integrated optics, higher power, and shorter wavelength semiconductor diode lasers, x-ray lasers, and so on. However, the understanding of quantum optics is advancing rapidly through experiments on such things as squeezed states and photon correlations, which make evident the strangely nonlocal nature of light and matter. Very possibly something quite strange and novel will come from this research.

Q: What are you working on now?

Schawlow: As I approach the university's compulsory retirement age of 70, which I will reach in 1991, I am trying to help my last graduate students finish their thesis research. We have been working on systematic laser labeling methods for analyzing molecular spectra. In another experiment, we are looking for very weak, sharp absorption lines in solids such as rare earth metals and thin films. So far this does not involve lasers at all, but I was doing spectroscopy years before there were any lasers. We are beginning to explore the uses of tunable diode lasers, and I expect to pursue that on a small scale for the next several years unless I can think of something better that is not very expensive.

An earlier version was published in *Lasers and Optronics* ®
(formerly Lasers and Applications) a Gordon Publications, Inc. publication.

NICOLAAS BLOEMBERGEN

Masers and Nonlinear Optics

Born in Dordrecht, the Netherlands, in 1920, Nicolaas Bloembergen came to Harvard University after World War II to complete his doctoral thesis research on nuclear magnetic resonance. He joined the Harvard faculty in 1951 and was named Gordon McKay Professor of Applied Physics in 1957. He later became the Rumford Professor of Physics, and in 1980 he was named Gerhard Gade University Professor at Harvard. He became Professor Emeritus in 1990.

In the 1950s Bloembergen pioneered development of the three-level solid-state maser. After the laser was developed, his research interests turn to nonlinear optics. His theoretical work, which laid the groundwork for the field, was first published in the early 1960s and summarized in his 1965 book, *Nonlinear Optics.* That work on nonlinear interactions also led to development of techniques for extremely high resolution laser spectroscopy. He received the 1981 Nobel Prize in Physics "for the development of laser spectroscopy," sharing it with fellow laser researcher Arthur L. Schawlow and with Kai Siegbahn of Uppsala University in Sweden, who pioneered electron spectroscopy.

Figure 4.1 Nicolaas Bloembergen examines a chromium-doped potassium-cobalt-cyanide crystal used in an early maser in March 1958 (courtesy of Harvard University).

Bloembergen's other honors include the 1961 Stuart Ballantine Medal from the Franklin Institute, the National Medal of Science in 1974, the Lorentz Medal of the Royal Netherlands Academy of Sciences in 1978, the Frederic Ives Medal from the Optical Society of America in 1979, and the IEEE Medal of Honor in 1983.

Jeff Hecht conducted this interview on November 5, 1984, in Bloembergen's office at Harvard. It was updated in May 1990.

Q: How did you become interested in science?

Bloembergen: In Holland I went to a Latin school, a "gymnasium," which included the equivalent of the first two years of college in America. The primary fields of study were the humanities, but we had very good science teachers with doctorates in mathematics, physics, chemistry, and biology. I became interested in physics as the most challenging subject, because it combined theory with experiment and mathematics with physical phenomena. I still feel that way.

Q: Your education must have been disrupted by World War II. What eventually brought you to the United States?

Bloembergen: They [the German occupation regime] closed the universities in 1943, and I had to hide and just try to survive until the end of the war. Afterwards I passed my qualifying [doctoral] exams in Holland, but circumstances were so difficult that I wanted to get out. I had always planned to do my thesis research somewhere else, and after the war the United States was the only place to go.

Q: You started out working in nuclear magnetic resonance. What took you to masers and eventually nonlinear optics?

Bloembergen: It was a very natural evolution. I wanted to branch out and learn something about microwaves. I was interested in nuclear resonance saturation, so I studied the same problems in microwave magnetic resonance. The maser came along as a very special application of saturation in magnetic resonance, where one field at a high frequency saturates the resonance and causes amplification at a lower frequency. The result is what is called an inverted or negative-temperature population. My colleagues [Edward M.] Purcell and [Robert V.] Pound here had discussed these things for NMR, but never with practical applications in mind.

Q: How did that lead to new approaches to masers?

Bloembergen: The ideas involving population inversions in NMR were only for transient cases. For the maser, I devised the three-level pumping scheme, which is a steady-state solution so you can get gain and maintain it afterwards. [Townes's first ammonia maser was a two-level system, where the inverted population was produced by separating excited ammonia molecules from those in lower states.] Three-level pumping originally involved microwaves, but clearly the scheme is very general, and it is the basis of essentially all lasers as well. Historically, that pumping scheme came first for microwave masers. Then Townes and Schawlow saw the opportunity to use it at optical frequencies.

Q: Did you use a resonant cavity when you tried to build a maser?

Bloembergen: Yes, that's sort of natural, though we also had a traveling-wave type. You really had to have two cavities, one resonant at the pump wavelength and one at the signal wavelength. We built one and it worked, but it wasn't the first one. We had been concentrating on a maser amplifier for the 21-centimeter wavelength of intergalactic hydrogen, which is the prime scientific application of the maser. Unfortunately, it turned out to be harder to get amplification at 21 cm than at the X band, and the first three-level maser was one at Bell Labs that amplified at the X band after being pumped with the K band.

Q: Your first solidstate maser was from a chromium-doped crystal of potassium cobalt cyanide. Didn't that lead to an interesting incident?

Bloembergen: In 1958, Charles Townes and I shared the Morris Liebman Award from what was then the Institute of Radio Engineers and has since become IEEE. We both went to a dinner in New York City and brought our wives along. Mrs. Townes talked to my wife and showed off a very nice pendant with a ruby crystal set in gold on a gold chain. Her husband had had it made for her in commemoration of the maser, which I think was a very nice gesture. That night in our hotel room my wife asked, "When are you going to give me something related to your maser?" So I said, "Well, dear, my maser works with cyanide."

Q: Did you try to build a laser?

Bloembergen: No. I was well aware that in principle you could adapt the maser scheme to higher frequencies, including the optical range. But it was too risky to try with the small group here. One reason it took us longer [than Bell Labs] to build the three-level maser was that we didn't have the support facilities here. I think all the early lasers were developed first at industrial research organizations. There was [Theodore] Maiman at Hughes with ruby, and [Ali] Javan at Bell Labs with the helium-neon laser. The carbon-dioxide laser was pioneered by Kumar Patel at Bell Labs, and the dye laser was conceived and built by [Peter] Sorokin at IBM. The semiconductor laser was at General Electric, IBM, and Lincoln Labs [a government-funded laboratory managed by the Massachusetts Institute of Technology]. What was really needed was a big support organization which could focus different technologies on a common goal. At American universities, you plod along with small groups of specialists. Harvard only provides buildings, but you really need support staff, technicians, glass blowers, opticians, and so on. History clearly shows who developed lasers first. You could try to duplicate them afterwards, but even that is hard.

Q: How did you get your first laser?

Bloembergen: We had to wait to buy the first model, from Trion in 1961. That's what everybody else [in the universities] did, and everybody who got one of those lasers discovered plenty of new effects. I saw that we were not going to discover any new lasers, but the least we could do was to use lasers to study optical properties at high light intensities. When Maiman realized the laser in 1960, we already were working on microwave modulation of light beams. Then I decided to do modulation of light with light, which was essentially nonlinear optics.

Q: Did you start out expecting to see nonlinear effects?

Bloembergen: Yes, but the first to see a nonlinear effect were Peter Franken and his group, who produced the second harmonic of light. It was very exciting. I was at the meeting where he announced it, and I said, "Peter, how did you do it?" Peter was very nice and modest, and said, "Nico, if you had one of those Trion lasers and you shot it at something, you would have done the same." I am not sure if I would have thought of it, but that's what happened.

In very quick succession [Robert] Terhune, [Joseph] Giordmaine, and others discovered lots of harmonic-generation properties. We had already been working on the theory when Franken published his experimental results. We already had formulas on paper. His work stimulated me, and after six months we had long theoretical papers ready, which were distributed in early 1962 and came out in *Physical Review* in April and June. We predicted reflected harmonics and so on. Reflection, transmission, polarization, refraction, and other optical effects all have analogs in the nonlinear regime, so we verified them.

Q: How did you come to concentrate on the theory of nonlinear optics?

Bloembergen: I always liked to do theory. Also, there are quite a few nonlinear effects in magnetic resonance, and essentially the maser pumping scheme is one of them. I was aware that a high field intensity at one frequency could modify properties at another frequency. We studied light modulation by microwave irradiation and wrote a paper on the subject, which is now mostly forgotten, but all the theoretical formalisms are very similar, so we applied them to optical frequencies.

The fields of lasers and nonlinear optics would not have developed so quickly if it hadn't been for the wide body of knowledge from microwave spectroscopy and magnetic resonance. Schawlow and Townes had experience with microwave spectroscopy of molecules and gases. I had my experience in magnetic resonance. It is part of a very logical historical evolution.

Q: The theory of nonlinear optics seems quite complex, with formulas that can run over a page long. Is it more so than NMR theory?

Bloembergen: No, NMR has developed in great detail, too. Now the basic principles in optics are well established, and we're digging deeper and finding more detail, to dot the i's and cross the t's.

Figure 4.2 Bloembergen celebrates receiving the 1981 Nobel Prize in Physics (courtesy of Harvard University).

Q: What sort of experimental work did you do?

Bloembergen: We got the lasers going and verified many of the predictions. The field got so popular you had to pick topics that graduate students could do without being scooped by much larger organizations, but we did quite a bit. We provided the first experimental verifications of the generalization of optical laws of reflection and refraction into the nonlinear regime. We also measured nonlinear susceptibility and its dispersion as a function of frequency. In many cases we were the first to study the phenomena in detail. Then we got into very high light intensities where even dielectrics break down and suffer optical damage. That has turned out to be very important for high-power lasers. When short picosecond pulses came along, we could do time-resolved spectroscopy and nonlinear studies. I found optics so interesting that I dropped the magnetic resonance stuff in the mid-1960s.

Q: What sort of technical problems did you encounter?

Bloembergen: I chose problems of an academic nature so the technology wouldn't become too tricky. I always liked small-scale experiments, because a single person can understand the details and individual measurement problems. I knew that unless we started a big program, spent big money, and got a big staff, we couldn't adequately solve problems that presented very big technological challenges.

Q: Did you ever wish for an industrial lab with more equipment?

Bloembergen: Yes, but I solved that by consulting and by talking to industrial scientists. Sometimes I was envious, but I always had some ideas we could pursue without big experiments.

A small organization has its advantages. The working conditions are good for small-scale experiments. Students here can talk directly to the machinist in the shop, where in a big organization they have to wait and go through channels. In the 1960s the climate for funding was very liberal. We had the Joint Services Electronics Program, which let us shift from magnetic resonance to optics without any explicit proposal. It was not until the late 1960s that I had to start writing proposals and all that.

Q: Who has worked with you in nonlinear optics?

Bloembergen: My coauthors are listed on all the papers. The early ones were John Armstrong, now at IBM; Peter Persham, now my colleague here at Harvard; and J. Ducuing, a Frenchman who was for a time director general of the French national research organization CNRS and now is chief scientist for NATO [the North Atlantic Treaty Organization]. There was Richard Chang, now at Yale, and Ron Shen, a professor at Berkeley who just published a book, *Principles of Nonlinear Optics.* It is the first really big improvement on my early book, which I am happy to say is not wrong, but far from complete. Two of my more recent students also have written books in the field recently, Marc Levenson [of the IBM San Jose Research Laboratory] on nonlinear spectroscopy and John Reintjes of the Naval Research Laboratory on nonlinear processes in liquids and gases.

I have had many good students here, including some other very remarkable people, and I don't want any of them to feel shortchanged. I have supervised a total of 58 PhD theses and about as many postdocs over the years. I couldn't have published as much without all those people.

On May 27, 1990, many of my former graduate students and postdoctoral associates assembled at Harvard University for a one-day scientific symposium. This was followed by a reception and dinner to celebrate my seventieth birthday and my retirement from Harvard. It was a memorable occasion, reuniting those who have worked over the past four decades in my research group in the Gordon McKay Laboratory of Applied Science.

Q: Were you surprised at the success of nonlinear optics?

Bloembergen: I'm still surprised. We can transform 80% of an infrared beam into the ultraviolet. And the enormous frequency resolution made possible by nonlinear spectroscopy, developed by people like Ted Hänsch, is fantastic.

Q: Had you expected that resolution theoretically?

Bloembergen: I never really took time to think about it, but I probably should have. I always concentrated on properties at high power, which in some cases is just the opposite of high resolution. But it isn't always. Very recently we found a line that is only 10 kilohertz wide, caused by collision-induced coherence. Everything goes just the opposite of what you might expect. The lines get both more intense and narrower the more collisions you have. Originally we predicted these collision-induced coherent effects in one of those page-long formulas you mentioned in my 1965 book. Nobody had ever seen that term then, and some people said it probably didn't pertain to reality, so I was very gratified that we demonstrated it experimentally a few years ago.

As always, if you have a new handle, you find some new things.

We have been studying these optical resonance effects in phase conjugation, and it is really amazing how fine the frequency features are. We are doing phase conjugation in sodium with four or more atmospheres of argon buffer gas, and it seems that the more buffer gas you put in, the better things get. We have to find the limits, so we are going to put in 10 or maybe 100 atmospheres of argon and see what happens. It's quite exciting; there always are surprises.

Q: Could there be other terms in your early equations waiting to surprise you?

Bloembergen: I don't think so, but some people say that if you go to higher-order nonlinear effects, such as those quartic in the field amplitudes, you still may see surprises. My colleague Tom Mossberg here at Harvard is getting new optical echoes. It is amazing how such things let you store information optically and perhaps recall it after quite a long time. Other groups are working on that also, so optical echoes still may have surprises.

Q: Where do you think nonlinear spectroscopy and nonlinear optics will go?

Bloembergen: Nonlinear optics will grow more and more into applications. Fiber-optic communications is one of the biggest ones, because clearly you have to know the nonlinear optical properties of the fiber. People can use the formulas we developed and analyze fiber properties in much more detail to build the best devices.

Medicine is probably the second most important application. Lasers can be used in very sophisticated ways in surgery. They even use short pulses for certain problems in the eye, to get optical breakdown at the lens without damaging the retina.

Scientifically, applications will be in ultra-short-pulse research; lots of phenomena have to be investigated in the femtosecond regime. There is an awful lot of photochemistry to be done. Basic physics may be last in importance except in high-resolution spectroscopy to check gravitational theory, and to measure some more decimal places in the interaction of hydrogen atoms to challenge the theorists.

Figure 4.3 Nicolaas Bloembergen in his Harvard office (courtesy of Harvard University).

Metalworking and materials-working clearly will grow substantially in the next decade. Every year the Materials Research Society meets here in Boston, and there is always a big laser component. Lasers will be used in geophysics to look for earthquake precursors and used in clocks to test special relativity.

Q: Aren't some of those areas more general laser applications than specific applications of nonlinear effects?

Bloembergen: They are laser applications, but to get high resolutions you always need nonlinear effects. To eliminate Doppler broadening you need either two-photon absorption or saturation, which is really a nonlinear effect. It's quite surprising how nonlinearities crop up everywhere.

I feel not surprised, but very pleased, that the wide spectrum of laser applications penetrate so many different fields of human endeavor. Now there is the idea of using high-energy lasers as defensive weapons.

Q: I was surprised to see you getting involved in that controversy, as co-chair with Kumar Patel of the American Physical Society's study group on "Star Wars," or the Strategic Defense Initiative. Did they have to twist your arm?

Bloembergen: Yes, I guess they twisted my arm. I didn't relish it, but I think it is a job that had to be done. In most studies so far, most of the attention has been paid to politics. They say, "Suppose we had an extremely powerful laser up there, what would happen?" Our study was different, restricted to assessing the proposed technology.

This study was completed in September 1986 and published as a special supplement in the *Reviews of Modern Physics*. It is now internationally accepted as an authoritative and scientifically correct description of the technology of directed energy weapons at the end of 1986. Since that date, the Strategic Defense Initiative Organization has used it as a yardstick to measure technological progress. It is noteworthy that the debate over deployment of directed energy weapons for strategic defense purposes has quieted down considerably since our report appeared. The more time elapses, the better I feel about the report. It has become part of the accepted archival literature.

Q: What has most surprised and impressed you about laser development so far?

Bloembergen: The growth to technological maturity of an annual laser market of several billion dollars. The commercial market now dominates the military by about a factor of three. Applications in optical communications, in surgery, and in other fields have an impact on the large segments of our society.

Q: What new directions do you expect laser technology to take in coming years?

Bloembergen: The importance of semiconductor lasers and integrated solid-state devices will grow substantially. A tiny semiconductor laser is incorporated in every CD [compact disc] player.

More powerful semiconductor laser arrays may replace gas lasers in many applications.

Q: What are you working on now?

Bloembergen: I am now officially retired, but in 1990 I am president-elect of the American Physical Society and I shall serve as its president in 1991. I hope to have time to keep up with exciting new developments by participating in conferences and by finding time to catch up with the research literature.

Q: If you had it to do all over again, would you still work in this field?

Bloembergen: Yes, it has been a good field, a very fruitful field. But my answer would be different if I were to start out as a young graduate student now. I wouldn't go into it because the field has matured. I probably would go into biophysics and biochemistry, because there you can still do some relatively small-scale experiments. If you are clever in your experimental techniques and understand the field, you can do some really clever things quickly in a small group.

An earlier version was published in *Lasers and Optronics* ®
(formerly Lasers and Applications) a Gordon Publications, Inc. publication.

GORDON GOULD

Another View of Laser Development

Gordon Gould's patent claims have made him one of the most controversial figures in the laser world. As a graduate student in Columbia University in his mid-30s, he was exposed to early maser work in the 1950s. Drawing on his earlier background in optics, he outlined a plan for trying to build a laser. Instead of trying to publish his results in a scientific journal, he chose to seek a patent.

His original patent applications went through many years of delays and many challenges, but he finally received four basic U.S. patents: one in 1977 on optically pumped (or excited) laser amplifiers, one in 1979 on a broad range of laser applications, one in 1987 on electric discharge pumped lasers, and one in 1988 on Brewster angle windows for lasers. The validity of these patents has been upheld in several court actions, and virtually the entire American laser industry has been licensed under them.

In 1973, Gould helped found Optelecom Inc. in Gaithersburg, Maryland, a maker of fiber-optic equipment and systems, where he became vice president and chief scientist. He retired in 1985,

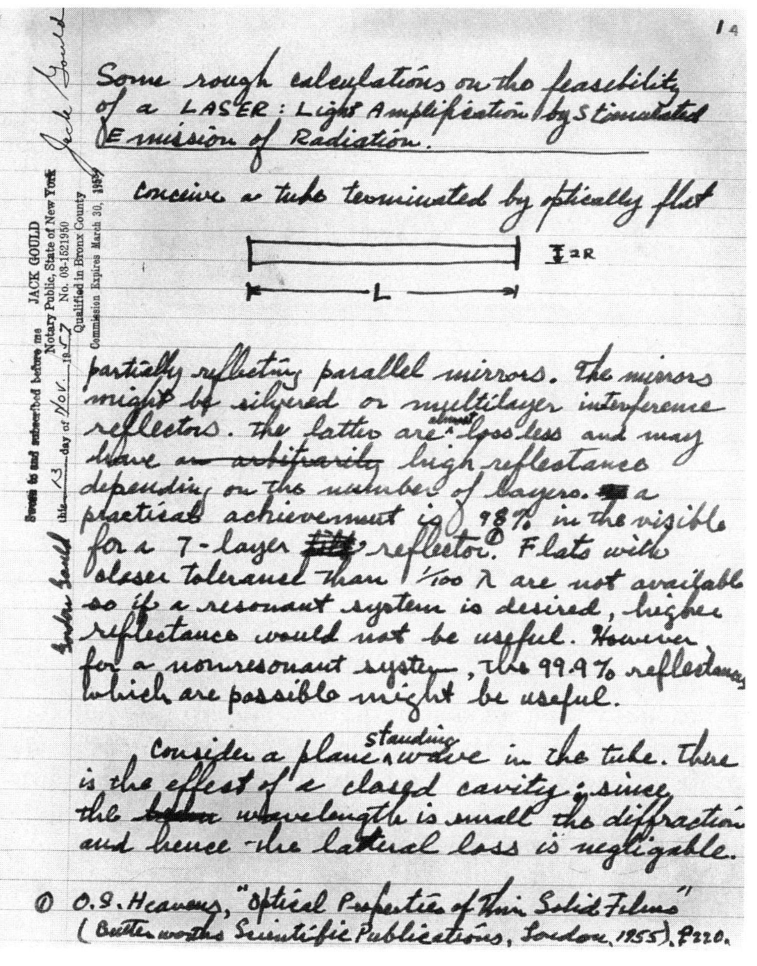

14

Some rough calculations on the feasibility of a LASER: Light Amplification by Stimulated Emission of Radiation.

Conceive a tube terminated by optically flat

partially reflecting parallel mirrors. the mirrors might be silvered or multilayer interference reflectors. the latter are almost lossless and may have ~~an arbitrarily~~ high reflectance depending on the number of layers. a practical achievement is 98%① in the visible for a 7-layer reflector. Flats with closer tolerance than 1/100 λ are not available so if a resonant system is desired, higher reflectance would not be useful. However for a nonresonant system, the 99.9% reflectances which are possible might be useful.

Consider a plane standing wave in the tube. there is the effect of a closed cavity; since the wavelength is small the diffraction and hence the lateral loss is negligable.

① O.S. Heavens, "Optical Properties of Thin Solid Films" (Butterworths Scientific Publications, London, 1955). P220.

Sworn to and subscribed before me this 13 day of Nov 1957

JACK GOULD
Notary Public, State of New York
No. 03-1521950
Qualified in Bronx County
Commission Expires March 30, 1959

Jack Gould

Gordon Gould

Figure 5.1 The first page of Gordon Gould's November 13, 1957, notebook describing his ideas for a laser (courtesy of Gordon Gould).

but remains a director with the company. He has been elected to the National Inventors' Hall of Fame.

Jeff Hecht conducted this interview on September 19, 1984, at Optelecom. It was updated in September 1990.

Q: How did you get involved in science and technology?

Gould: I always knew that I wanted to be an inventor, even before high school. I followed a course which led me in that direction, taking physics in college, then actually trying to invent. After a while, in the late 1940s, I realized that I didn't know enough yet to do the kind of things I wanted to do. For example, I had an artificial diamond project that produced a lot of graphite, but no diamonds, because I was too ignorant of the thermodynamics involved. That was when I decided I had to go back to graduate school, and did so at Columbia University.

Q: How did you get interested in optics?

Gould: It was at Union College, where I had a professor, Frank Studer, whose undergraduate optics course just fascinated me. After graduating in 1941, I worked a summer at Western Electric, which convinced me I did not want to climb all the rungs in that big an organization. I went on to graduate school at Yale, which is very well known for spectroscopy and optics. Then came World War II, and in 1943 I asked my professor, W.W. Watson, what he might suggest in the way of essential work to avoid the draft. He told me to go to a certain address in Manhattan and tell them he had sent me. I did that the day after my draft physical, and it turned out to be the Manhattan Project. Later, when I went to Columbia, my PhD thesis on optical pumping of a thallium atomic beam got me involved with optics again, and before I finished that project I had thought of the laser.

Q: How did Charles H. Townes's maser work at Columbia influence you?

Gould: It was very exciting, and obviously it got me to thinking along those lines, although my research project was not on a maser. I did think of an optically pumped maser, which incidentally I wrote down in a notebook and Townes witnessed. About a year later, I realized how to make a laser: a Fabry-Perot resonator would solve the problems of needing a low-loss cavity resonator for the laser light while at the same time allowing pump light to shine in to excite the medium optically. It didn't have to completely enclose the resonant light modes. I got that electrifying idea in November, 1957. About two months later, I got to thinking about this thing and knew that it was going to be the most important thing I ever got involved with in my life. I realized I would have to leave Columbia to work on it, because Professor [Polykarp] Kusch, my thesis advisor, would never let me substitute a thing like that for a research project of the very pure and basic type that was characteristic of Columbia, although Townes dabbled in such things.

Q: A theoretical description of the laser was not pure enough?

Gould: No, there should be no practical applications. That makes it pure right off. That attitude does not exist so much today, but in those days there was a very sharp distinction between basic physics and applied physics. Columbia did not deal with applied physics.

Q: Where had you picked up the idea of optical pumping?

Gould: [Prof. I.I.] Rabi came back from a conference in France, where the idea was introduced, very excited by the possibility of using optical pumping along with molecular or atomic beams—which were the big thing at Columbia—to make some new kinds of measurements. I had been working on the thallium beam for three or four years already, without much success, trying to excite the atoms thermally or by electric discharge. But I never got enough into the metastable state that I was trying to measure. Rabi said to try optical pumping, and for a graduate student that was an order. I tried it and it worked, and I was able to get something like 5% of the atoms into that metastable state. That introduced the idea into my mind, and it fermented there. First I came up with the optically pumped maser and then the laser. But it wasn't so much optical pumping that was the exciting thing, it was the concept of the resonator. Optical pumping of a gas was just the first of several different kinds of excitation I thought of for laser media.

Q: How did you get the resonator idea?

Gould: While I was at Yale, I used Fabry-Perot resonators and became familiar with the tools of optical spectroscopy. Years later I went to Columbia, which was big on microwave spectroscopy. To think of the Fabry-Perot as a resonator for a laser oscillator I had to have both those kinds of experience. It just clicked that one exciting night, about one in the morning, and I jumped up and started writing, and wrote that whole first notebook in one weekend. Then I had it notarized on Monday.

Q: Was that all night, all weekend?

Gould: Yes, which I could do in those days. Now in the case of Townes and [Arthur L.] Schawlow, who thought of this independently several months later, the combination of experiences was in two different people. Schawlow did his thesis in optics at the University of Toronto; Townes was the father of microwave spectroscopy and the maser. People ask, "Did it really hit you like a bombshell right out of the clear blue?" Well, in a certain sense it did. It also involved 20 years of stuffing necessary bricks and mortar into my mind for a purpose I didn't know.

Q: Once you had outlined your ideas for a laser, you tried to build one. What were the problems that kept you from having the first working laser?

Gould: The first one was money. Realizing that I couldn't do the work at Columbia, I left in 1958 to work at Technical Research Group Inc., TRG, out on Long Island. I told Lawrence Goldmuntz, the president, that I had some ideas which I would like to reserve out of the usual patent agreement one signs when one comes into a company. He said okay, write them out and we'll exclude them. I hadn't written up the laser very well, and I had to work on other things as well. By the time I was done it was clear that some of the work had been done while I was with TRG. Eventually we agreed on a split of rights; I would retain certain rights and TRG would also have some. Goldmuntz became excited about the project, so TRG began looking for research support.

Figure 5.2 Gordon Gould in his laboratory in the 1960s (courtesy of Gordon Gould).

Q: Where did you finally get your money?

Gould: From DARPA [the Defense Advanced Research Projects Agency].

Q: We have heard that you asked for $300,000 and they gave you a million. That's something that normally does not happen.

Gould: That's for sure. But they were exceedingly happy about the prospect of a death ray, Buck Rogers style, although I wasn't so hot on that idea. There were plenty of real applications for the laser, but that's what got them in. There was a second problem that was still not a technical problem: the project became classified and I couldn't work on it after having gone to all that trouble. I was considered a security risk because in the late 1940s I had been in a Marxist study group, so I could not get a clearance. The technical problems in building a laser were several, but I was not really working on the project, so I couldn't deal with those.

Q: How did you interact with the people working on the classified project at TRG?

Gould: I was assigned other tasks, including several other projects which were not on the subject of lasers at all. As far as the laser project was concerned, I was there as a consultant. If anyone wanted to consult with me, they could, without telling me what they were doing. I had a fair idea of what Steve Jacobs and Paul Rabinowitz were doing with an optically pumped gas laser. But the real technical problems came because people working on the program didn't really follow my proposal at all, but set out to do other things instead of making lasers. For example, they started a great big project to learn how to grow crystals instead of taking a natural crystal and simply making a laser from it. The real reason why the first lasers were not built at TRG was simply because they didn't try.

Q: Eventually they did build one of the first lasers, didn't they?

Gould: Yes, the optically pumped cesium laser, pumped by helium emission. The only one built anywhere was at TRG.

Q: Whatever happened to the cesium laser?

Gould: It didn't put out very much power—about one-tenth the power of helium-neon—and the wavelengths were long, 3 and 7 micrometers. It also entailed massive problems with cesium, an alkali metal of the worst sort, so it would have been expensive to make. HeNe was much more practical, so nobody really gave it much thought.

Q: Were you at all surprised when the laser worked?

Gould: Not at all; I knew it was going to work.

Q: Were you surprised that Ted Maiman was the one who came up with it?

Gould: I was surprised at that; I didn't even know he was working on it. He sort of startled the world. Schawlow at Bell Labs and Irwin Wieder at Westinghouse had made some measurements that seemed to show that ruby fluorescence was very inefficient, but they failed to account for self-absorption, which was happening with a vengeance in ruby. I believe Maiman ground up some ruby, so the distance light had to travel to the surface was very small; it got more than 95% quantum efficiency, so he knew he had enough fluorescence to make that laser work. A group at TRG had started to work on ruby, which I had suggested among other things. Then they listened to a talk by Schawlow, who turned everybody off of ruby, and they put it on the shelf. After Maiman showed that it did work, they made one within one month at TRG. Similarly at Bell Labs they had one working within a month. They both heard about it through the grapevine before it was published.

Q: Your patent application spelled out a number of possibilities for laser action. Was TRG trying them all?

Gould: No, even the million dollars was not enough for that, but they did select about six things. One was the potassium-emission-pumping-potassium scheme that Townes suggested, which has never worked. For seven or eight months they ignored cesium, and it was only a great effort on my part that got the projects shifted around. That might have been the first gas laser if they had started working on it right at the beginning. They also worked on a sodium-mercury discharge laser, which I had spent a lot more time discussing in my proposal than HeNe. They never succeeded in building it. For years, people opposing my patent application on the discharge laser cited the fact that nobody had built a sodium laser. Recently we decided to scotch that line of reasoning by building one. It is sitting right out there in the lab [at Optelecom] now, and it does work, just like I said it would. The people involved in my efforts to get a patent on the discharge-pumped laser came up to see it and were astonished.

Q: Did you develop any other types of lasers?

Gould: The copper-vapor laser, in work that started at TRG, which I took to the Brooklyn Polytechnic Institute when I moved there in 1967, as an instantly tenured full professor. It is an interesting and unique type of laser, which after many years of trying to solve technological problems is now on the market. Eventually it will be more important for high-power applications than either ruby or YAG lasers. Those lasers have a fraction of 1% efficiency, so to get a kilowatt you need a power supply that fills up a room. Copper vapor has better than 2% efficiency. However, it is hard to make into a product because you have to heat it to 1700° to vaporize the metal, and that takes a pretty special type of furnace and costs a lot of current. What makes it possible now is that someone designed it so the energy used to excite the discharge also heats it to the right temperature.

Q: Do you have a patent on copper vapor?

Gould: Yes, but the market is too small to bother with now. A market has to be $20 to $30 million a year before it's worth trying to force people to license a patent, because that's going to cost you half a million dollars, even if your patent is iron-clad.

Q: Whatever happened to the TRG interest in your patents?

Gould: In 1965 TRG merged with Control Data Corporation and was put into an aerospace division. It lost money from the very day it merged into CDC though it had always made a profit before. In 1970 they liquidated it, but because of a clause in my patent agreement, they couldn't just turn around and sell the patents to someone else. Hadron bought the laser business but didn't want to spend the money needed to defend the patent claims, so the Control Data vice president handling the liquidation asked me to make an offer. I didn't have much money, but my lawyer suggested I offer them a dollar and 10% of whatever I made from my patents, and I got them. That was the first time I really owned them, and boy did they start burning holes in my pocket. Within three years I was out of money and trying to sell them. At that point I met up with Refac, who bought in lawyers who had been very successful before, and they started producing results, getting the first two patents [on optical pumping and laser applications] issued.

Q: Although your patent claims have been disputed, aren't you generally credited with coining the word "laser" in your notebooks?

Gould: Yes, and there is another story I would like to tell about that. I was one of the early presidents of LIA, back when the initials stood for the Laser Industry Association [it is now the Laser Institute of America]. The Electronics Industries Association, EIA, didn't like that because it sounded too much like their name, and because LIA was trying to work with EIA's laser group, we tried to find a different name to accommodate them. We were thrashing around the name question at a board meeting out in Arizona when Arthur Schawlow looked up with a twinkle in his eyes and said, "I know what to call it: the Optical Maser Association." Then Peter Franken said, "You lost that battle long ago." [Townes and Schawlow had used the term "optical maser" in their early papers.]

Q: Your optically pumped laser patent was issued in 1977, some 18 years after your application was filed. What caused the further 10-year delay before your last two patents were issued?

Gould: In 1982 the Patent and Trademark Office started a reexamination of the 1977 and 1979 patents, at the request of General Motors, AT&T, and the Control Laser Corporation. GM came to the Patent Office with a foot-thick pile of alleged "prior art," including such citations as Archimedes's use of focused sunlight to start fires, as anticipations of the laser. These reexaminations dragged on four more years. Meanwhile, the electric-discharge pumped laser patent was slowly winding its labyrinthine way through to its final (and expected) rejection by the Patent Office examiner. It was only after these cases were brought out of the Patent Office's internal administrative system and into the regular U.S. court system that judges and juries with no particular preconceptions could decide the issues based on the evidence alone. In no less than three full-scale trials in three different U.S. District Courts, the patents were found valid. The Patent Office finally issued certificates of validity for the two reexamined patents and issued the last two patents in 1987 and 1988. Since then, some 200 laser manufacturers have willingly signed royalty-bearing licenses.

Figure 5.3 Gordon Gould in 1990 (courtesy of Gordon Gould).

Q: How did you come to be vice president and chief scientist at Optelecom?

Gould: By 1973 the pressures on professors like me at a place like Brooklyn Poly were getting very bad. More students, more classes to teach, and at the same time more demands or research contracts to support the graduate students. Bill Culver as trying to get me to join him to start Optelecom. Meanwhile, Brooklyn Poly had merged with the engineering department of New York University, which was losing money hand over fist because the number of students had been declining for several years and there were too many tenured professors. So the management proposed to give anybody who would resign his tenure a whole year's terminal leave with pay. Since I was going to quit anyway, I jumped at that. That year, which I worked without pay, was my investment in Optelecom.

Q: What technology were you developing at Optelecom?

Gould: Before I retired in 1985, I had moved far away from lasers themselves, except for building the little sodium-mercury laser back in the lab. At this point, I was in an optical communications company. I picked out a special niche so we didn't have to try to compete with the big telephone companies. That niche is developing specialized cable for oil-well logging, where you send an instrument package down a well hole and measure what is there. That can be 30,000 feet down, where the temperature is 500°F. Traditionally, they have used an armored electrical logging cable, which is limited to 70,000 bits per second. That sounds like a lot, but it's not enough to carry the data that these instruments can generate, so there is pressure to get an optical fiber into that cable. We designed a new type of fiber-optic cable and have delivered a complete telecommunication system based on it to Chevron, and they are using it. It may seem far afield from lasers, but actually we use a laser transmitter in our systems.

Q: Is that because of the distance or the data rate?

Gould: The conditions at the bottom of the hole are too extreme for a laser. So we have the laser at the top of the hole, send the light down, modulate it down there, and send it back up. We use a Nd:YAG laser that puts out a watt or so. By the time we get to the bottom of the hole, it's down to a milliwatt, and by the time we get back up again, it's down to a microwatt, but that's still plenty.

Q: Does the present state of laser technology surprise you?

Gould: The only laser that did surprise me was the semiconductor laser; it surprised me that it was feasible. Other laser developments, even though I may not have thought of them, didn't surprise me; they seemed rather natural when they came along, like the copper laser. I particularly am not surprised by the use of really high-powered lasers for industrial materials processing because I envisioned doing that right from the beginning in my patent application.

Q: Where do you think it's going?

Gould: Way back in the beginning, I used to say that lasers are to light technology what vacuum tubes or transistors were to electronics. We certainly have moved in that direction, but we haven't gotten there yet. One day the field that you might call optronics may be as big as the whole electronics industry is today. The laser industry passed a billion dollars a year a while back, and it will continue to grow in volume and in the number of different kinds of applications, which will become increasingly pervasive.

Q: Have any applications surprised you, say laser fusion?

Gould: No, because in my very first notebook I wrote: "The powers that will be available with the media, that I can see are feasible, will be able to heat an object up to 100 million degrees." So I wrote down the possibility of starting nuclear fusion. They've been working on it for 25 years and it's not here yet. But it doesn't surprise me.

Q: What do you think about laser weapons?

Gould: You remember the reason that DARPA got so excited was because of the concept of laser weapons. What has not surprised me is the fact that they don't exist yet, or they're not operational yet. My reaction to the death-ray idea was that light is not really suitable because its wavelengths are too long. Diffraction spreads light waves out too fast for them to be effective at any great distance, and really you should be using much shorter wavelengths so the diffraction is negligible. In fact, how about matter waves? How about a bullet? That was my reaction way back in the beginning, and it is still my reaction. I don't think that much of the "Star Wars" concept. It is particularly dangerous to get people's expectations up beyond what may be feasible. They'll think that the security of the country is being improved when it really isn't. It may even drive the Russians to do things neither they nor we would want them to do.

Q: If you had it all to do over again, would you still work on lasers?

Gould: I certainly would, but I would do it differently than I did before.

Q: Would you look for a different lawyer than the one to whom you first showed your notebooks, and asked how you should go about getting a patent?

Gould: That would be the first thing. Just think, if I had understood him and if he and I had communicated properly in January 1958, this whole history would have been entirely different. I would have had my patent long, long ago, and it would have run out long, long ago. I would have made, maybe, $100,000, much less than the patent has brought me now. But certainly the laser proved to be what I realized it was going to be, the most important thing I ever got involved with in my life, and if it happened to me again, I would do it. Of course, at that moment in my life I was too ignorant in business law to be able to do it right, and if I did it over again probably the same damn thing would happen.

An earlier version was published in *Lasers and Optronics* ®
(formerly Lasers and Applications) a Gordon Publications, Inc. publication.

THEODORE H. MAIMAN

The First Laser

Born in 1927, Theodore H. Maiman earned a bachelor's degree in engineering physics from the University of Colorado and a master's in electrical engineering from Stanford University. Like many other laser pioneers, he studied microwave spectroscopy, writing a dissertation on the topic under Nobel laureate Willis Lamb at Stanford to earn a PhD in physics in 1955. After briefly working at Lockheed, he joined the staff of Hughes Research Laboratories, where he worked on maser development.

Maiman's maser projects at Hughes included the first ruby masers cooled with liquid nitrogen and dry ice. Like many others, he became interested in the possibility of building lasers at visible and infrared wavelengths. He studied several solid-state materials, eventually settling on ruby as the most promising. Undaunted by claims that ruby would not work. he pressed ahead, working without management support. On May 16, 1960, when he was 33 years old, he demonstrated the first laser.

That laser marked a dramatic breakthrough. Other researchers were caught in conceptual logjams, trying to obtain laser action

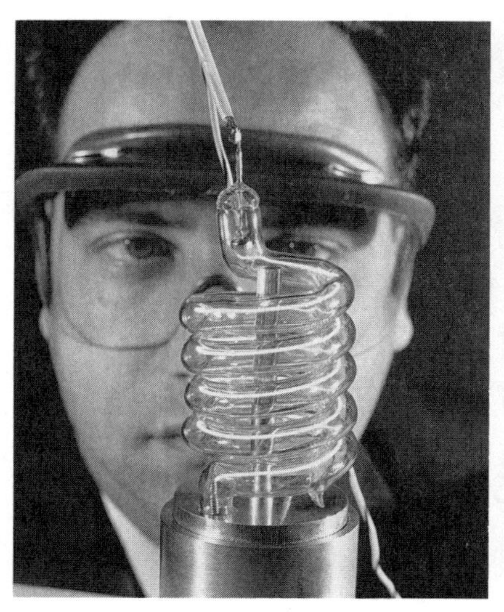

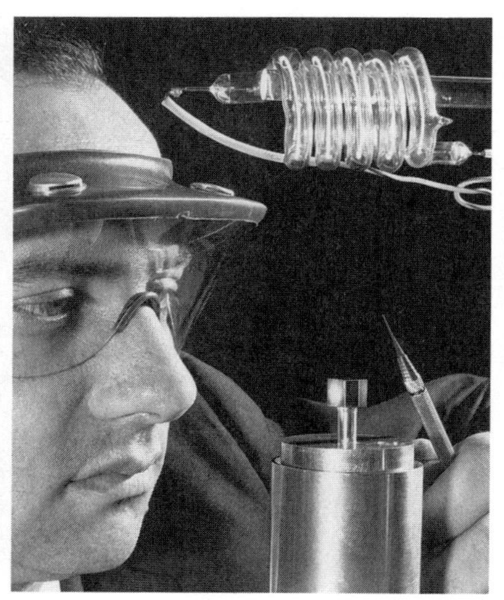

Figure 6.1 (a) One of the first ruby lasers, a ruby rod surrounded by a spring-shaped flashlamp, with Theodore H. Maiman in the background. (b) Theodore Maiman examines a ruby crystal at Hughes in 1960, with a ruby laser rod and flashlamp hanging in the foreground. These photos were widely distributed, and some groups built lasers using the flashlamp type shown—but Maiman used a smaller, less photogenic helical lamp in his first laser experiments (courtesy of Hughes Aircraft).

from such difficult media as alkali metal vapors. Maiman had the stroke of genius needed to take a different approach. His design was of elegant simplicity: a small ruby rod, with mirrors coated on its ends, inserted in a helical flashlamp. A bright pump pulse brought forth a short laser pulse. If his invention seems obvious to some today, it was far from obvious in 1960.

Dr. Maiman declined to have an interview published in this book. The narrative that follows is drawn from published sources cited at the end of the chapter.

Theodore Maiman became acquainted with microwave and optical research while a graduate student at Stanford, where he devised a way to analyze the excited energy states of helium using optical, electronic, and microwave instruments. This experience proved to be useful when, in 1956, shortly after the ammonia maser was developed, he joined the new atomic physics department at the Hughes Research Laboratories, then in a Hughes Aircraft plant in Culver City, California. The primary goal of the department was to develop new sources of coherent radiation at frequencies higher than were then available.

Maiman's first project at Hughes was to generate harmonics using a nonlinear cyclotron resonance under a basic research contract from the Army Signal Corps. That work was phased down, however, when the Army decided it would rather have a maser operating in the microwave X-band, 5.2 to 10.9 gigahertz. The X-band maser project, begun in 1958, was not particularly ambitious. Maiman was put in charge, although he said he "wasn't too enthusiastic at first because that was a practical device and I was more research-oriented" (Cavuoto, 1985, p. 86).

Military research agencies were intrigued by solid-state three-level masers because their extremely low noise levels made them attractive for radar communication systems. Their design, however, posed serious practical difficulties. One problem was their operating temperature of a few kelvin, requiring a double dewar with both an inner flask of liquid helium at 4 K cooling the maser and an outer flask of liquid nitrogen at 77 K to cool the liquid helium. Another was the need for an external magnetic field to create the proper conditions within the solid-state rod.

Early maser developers tested several solid-state materials, but most settled on ruby because it was rugged and easily grown. It also was the material used at Hughes for the Army contract. Maiman made several key innovations to improve the performance of ruby masers. One was to paint the ruby crystal with highly conductive silver paint to form a microwave resonator, leaving a small hole for the transfer of the microwave energy out of the cavity. The other was to change the magnet configuration.

Conventional wisdom at the time held that permanent magnets did not belong inside dewars because they might crack and interfere with the active crystals. Dewars containing masers thus had to be placed between the poles of large permanent magnets weighing about 5000 pounds. Naturally, this rendered masers far too cumbersome for practical use. Maiman showed that a magnet could be placed inside the dewar, allowing a much more compact design. His entire maser, including the dewar and magnet, weighed only 25 pounds. It also was more stable then earlier masers and had a gain-bandwidth product ten times higher.

Maiman continued working to improve the ruby maser and soon built one weighing under four pounds. He also succeeded in making solid-state masers that could operate at higher temperatures—first at the 77 K temperature of liquid nitrogen, then at the 195 K temperature of dry ice. Although these improvements were important, they were not enough to make masers into practical microwave amplifiers for most applications, and eventually other technologies became dominant.

Like many other maser researchers, Maiman was among the roughly 180 participants at the first International Quantum Electronics Conference, held in September 1959 at a resort in the Catskill mountains in southeastern New York. He gave a paper on ruby-maser physics, but had already begun studying the prospects for making masers at millimeter and infrared wavelengths. At that time he was thinking of optically pumping crystals doped with gadolinium ions (Bromberg, 1991, p. 86).

Most proposals for short-wavelength masers involved gases or metal vapors, but Maiman was wary of such media. The alkali metal vapors proposed by Townes and Schawlow were corrosive and difficult to use. Even the more benign inert gas systems required complex vacuum systems and were vulnerable to contamination. Presumably influenced by his maser experience, he concentrated on solid-state materials. After dealing with the frustrations of microwave masers, he wanted an optical laser material that would be rugged and not require cryogenic cooling. He also hoped to make a device much shorter than the meter-long cavity which gas-laser proponents thought would be needed.

Maiman initially considered ruby, but his first detailed calculations showed that it would be difficult to make a ruby laser. He was also discouraged by Irwin Wieder's measurements showing that only about 1% of the pump energy absorbed by ruby was transferred to the energy states needed to power an optical laser. This led him to turn his attention to gadolinium-doped crystals, because their sharp fluorescence lines made them seem promising as laser materials. Closer investigation, however, showed that the absorption lines also were narrow—making pumping difficult, especially because ultraviolet wavelengths were required.

At the September conference Maiman queried Wieder about details of his measurements. The answers led Maiman to suspect that ruby could be pumped with higher efficiency than Wieder thought possible. He also doubted Arthur Schawlow's statement that ruby was a poor laser candidate because it would be extremely difficult to excite half of the ions above their ground-state energy level. He began a thorough investigation of ruby, but his initial hope was to find out what was wrong so he could select a better material. His effort was limited, involving only himself and Irnee J. D'Haenens, his research assistant, and was supported by Hughes funds for internal research.

Maiman developed a detailed model of ruby, but he could not find any reason for it to have low efficiency. Instead, his measurements showed that the quantum efficiency should be close to 75%. This stimulated him to take a closer look and make more detailed measurements, which revealed that the quantum efficiency was nearer 100%. Wieder had been wrong.

Ruby also had other attractions besides high pump efficiency. A ruby laser would be small and operate at room temperature—without the cryogenics that impeded maser operation. Its emission was visible, unlike other proposed lasers which operated in the infrared, and Maiman has said that he was intrigued by the prospect of being able to see the beam (Cavuoto, 1985, p. 87).

The high quantum efficiency of ruby was encouraging, but it did not overcome all the remaining problems. Ruby is a three-level laser, with the lower laser level being the ground state of the light-emitting chromium ion. This meant that producing a population inversion would require depleting the ground state population. Maiman realized that would be a difficult problem, so he considered carefully the optical source needed to pump the ruby. His analysis showed that the critical parameter was brightness—a high enough photon flux could depopulate the ground state.

This did not solve the problem. His calculations showed that he needed a light source with a blackbody temperature around 5000 K. That would barely be possible with the commercial mercury arc lamps then available. Pumping a ruby rod would have required arranging mirrors, the pump lamp, and the ruby rod in an elliptical cylinder for the light to make multiple passes through the rod. Even then, laser operation would have been marginal.

Maiman then took what proved to be the crucial step of turning his attention to pulsed pumping rather than the continuous operation which virtually all other researchers were considering. Xenon flashlamps offered a much higher effective temperature than mercury arc lamps—about 8000 K. Also, he recalled later, "I didn't see any reason why I had to do this continuous-wave—pulsed mode was perfectly fine. People do a lot of things purposely with pulses, for example, radar. Besides, I was just trying to demonstrate that this could be done, not find the ultimate system" (Cavuoto, pp. 86–87).

After deciding to try to build a pulsed laser, Maiman pored through catalogs of laboratory light sources and calculated their effective brightness. Only three commercial lamps met his requirements, all helical quartz-xenon photographic flashlamps made by General Electric. He bought several of each: the small FT506, the slightly larger FT503, and the large FT624, which was made for aerial photography. He checked their ability to depopulate the ground state by monitoring an 11-GHz absorption of ruby when it was pumped by a xenon lamp. A decrease in absorption told him that the ground state was being depleted—encouraging him to continue working on the laser. First, however, he had to endure the move of Hughes Research Labs from Culver City to a new building in Malibu. He was back in the lab within two weeks after the disruption (Smith, 1984, p. 578).

The helical shape of the lamps initially posed a problem. Maiman had designed an elliptical cavity for optical pumping of a laser with a linear lamp. Helical lamps, however, did not fit well into the design. While he was pondering the problem, the lamp salesman told him that the largest of the helical bulbs produced light so intense that it could ignite steel wool. This led Maiman to realize that his focusing optics could not make the light any brighter than the lamp itself. With this realization, he decided to put the laser rod close to the lamp—inside its coil. He did put a reflector around the outside of the lamp, both to act as a radiation shield and to focus extra pump light back onto the rod.

To make a resonator for the ruby rod, he used an approach similar to the one he had taken with the microwave maser. He applied silver mirrors to both ends, coating the silver rather than painting it to avoid tarnishing. Initially he applied a totally reflective coating on one end and a partially transparent one on the

Figure 6.2 Theodore H. Maiman, at the time he received the Japan prize (courtesy of Corning Inc.).

other. However, the partial reflector was so thin that it tarnished easily, so he soon shifted to a total reflector with a coupling hole.

His first successful experiment, May 16, 1960 (Hecht, 1985), was with a ruby rod one centimeter in diameter and two centimeters long, which just filled the spiral of the smallest lamp. The entire package of lamp, reflector, and laser rod could easily be held in one hand, although it required a separate pulsed high-voltage power supply which was far more massive. In many ways that small—almost elegant—package foreshadowed the modern emphasis on miniaturization and compact solid-state lasers. However, in 1960 the marketing buzzwords were "bigger and better," and for many years most lasers would be much bigger than Maiman's first.

Maiman estimated that Hughes spent only about $50,000 on his nine-month laser research program (Cavuoto, 1985, p. 88), far less than Bell Labs spent on Javan's research or the government spent on TRG's efforts to build lasers based on Gordon Gould's proposal. Despite the small scale of his program, he felt opposition from Hughes managers, who apparently were swayed by Schawlow's arguments that ruby would not work. All that changed when he and D'Haenens detected clear signs of lasing.

The next few weeks were hectic ones for Maiman. Hughes management quickly lined up to support him. While his device produced the deep red pulses he had expected from a laser, it did not display the sharp threshold that had been expected for laser oscillation. The ruby rod was left over from earlier maser experiments, and its optical quality was limited, so Maiman ordered new ruby crystals so he could perform further experiments (Bromberg, 1991, p. 91).

Meanwhile, he felt intense pressure to publish his results before someone else made a laser. He feared that publication of his measurements of ruby fluorescence—showing high quantum efficiency—in the June issue of *Physical Review Letters* would put other researchers on the trail of ruby. On June 24 he submitted a paper describing the ruby laser to *Physical Review Letters.*

Unfortunately, Maiman's paper ran afoul of Samuel Goudsmit, an eminent physicist best known for his co-discovery of electron spin, who had founded the letter journal in 1958 to facilitate rapid publication of new results. Goudsmit had grown impatient with a glut of maser research, and in 1959 announced that the journal would not publish maser papers that did not "contain significant contributions to basic physics." He also opposed publishing series of articles from the same scientists—and Maiman had just reported on ruby fluorescence (Bromberg, 1991, p. 268, note 79). These policies led *Physical Review Letters* to reject summarily Maiman's paper, titled "Optical maser action in ruby," apparently without realizing its importance.

Maiman hastily dispatched a very short paper to the British weekly journal *Nature*, which published it on August 6. Hughes called a press conference on July 7 to announce the discovery, which got widespread attention in the national press. Hughes still distributes photographs of Maiman it took for the occasion, but they do not actually show the first laser. The photographer thought that the laser was too small for good pictures, so he substituted a later one which Maiman had built with the larger FT503 flashlamp. Because those photos were for a time the only information available on the pump light, scientists seeking to duplicate Maiman's laser went out and bought that lamp.

Meanwhile, Maiman suffered another publication indignity. An article he had submitted to the *Journal of Applied Physics* was obtained by the magazine *British Communications and Electronics*, which published it without authorization in its September 1960 issue.

Most other contestants in the race to build the first laser were surprised at Maiman's victory, and some hesitated to accept it. Maiman was one of the few developers on the west coast (today the site of extensive laser development and home to much of the laser industry). Like some other groups, he had revealed little about his work. Although it was long before the cold fusion fiasco, many American researchers were skeptical of publication by press release, particularly after public proclamations by Schawlow that ruby would not work. The skepticism was fuelled by the sketchiness of Maiman's intentionally brief paper in *Nature*. (He reportedly kept the paper short to assure rapid publication and leave open the possibility of publishing more details elsewhere.) In August a group at Bell Labs made a near-replica of Maiman's laser and demonstrated that it worked. They made careful measurements and published a detailed report which appeared in the October 1 issue of *Physical Review Letters* (Bromberg, 1991, pp. 91–92). This sequence of publications led to mistaken reports that Bell Labs had made the first ruby laser.

The ruby laser opened the floodgates to a tide of new lasers. At the IBM Watson Research Laboratory, Peter Sorokin and Mirek Stevenson had their doped crystals cut into rods, coated the ends, and put them in flashlamps to make the second and third lasers, as described by Sorokin in the next chapter. Interestingly—despite the earlier publication of the Bell Labs ruby laser report—in late November Stevenson and Sorokin still felt compelled to hand-deliver their paper to Goudsmit at *Physical Review Letters* and make a case for its publication.

Maiman's effort to build a practical laser—and avoid the problems already starting to choke off maser development—was eminently successful. Ruby lasers proliferated rapidly, as other groups duplicated his work. The sheer elegance and simplicity of his design belies the intellectual achievement it represents.

Maiman himself has labelled "incredulous" statements by others that the first laser was "easy" to make. He observed, "If it was so easy, why didn't Columbia, Bell Labs, or TRG pull it off? They each had a head start, plenty of money, and heavy staffing" (Cavuoto, 1985, p. 88). Like many other inventions that involve conceptual leaps, the laser appears simple only in hindsight.

It did not take long for ruby lasers to attract military development support and become commercial products. Hughes, as a leading military contractor, had two contracts for laser radar development by the end of 1960. Those programs, in turn, stimulated the development of commercial ruby lasers, which Hughes introduced in 1962 (Bromberg, 1991, pp. 126–128).

Other companies began making ruby lasers in 1961. One of the first was Trion, a spinoff of the Willow Run Research Laboratory of the University of Michigan in Ann Arbor. Maiman left Hughes in April 1961 to join Quantatron, a new company in Santa Monica which was trying to develop commercial ruby lasers and other high-technology products. When that venture ran out of money, Maiman and several coworkers founded Korad Inc. in Santa Monica in November 1962. The company was funded by Union Carbide, which retained 80% ownership; the remaining 20% was shared among Maiman and 10 others. Maiman was president and had the largest share.

Korad pursued both commercial products and military development. It grew to over $5 million in annual sales and over 100 employees in 1967. At that point Union Carbide exercised its contractual right to buy out the interests of Maiman and the others, and Maiman and several others left. Korad's sales shrank while it was part of Union Carbide, slipping to about $3 million in 1973, when it was acquired by Hadron Inc. (Laser Focus, 1977). The two companies merged their efforts, with Hadron moving from Long Island to Korad's Santa Monica facility. Interestingly, one of Hadron's previous acquisitions was TRG—where Gordon Gould had worked during the great laser race—which had earlier been purchased by the Control Data Corp.

Maiman was a consultant for several years after leaving Korad, before joining TRW Inc. as vice president of advanced technology for its electronics and defense divisions from 1976 to 1983. He was a founder of PlessCor Optronics Inc. (later renamed PCO Inc.), and remained a director until the company ceased operations in late 1991.

As the builder of the first laser, Theodore H. Maiman played a pivotal role in laser history. While the dispute over how to allocate credit for inventing the laser concept may never be settled satisfactorily, Maiman clearly built the first laser. That achievement has not uniformly brought him recognition. In October 1960, just months after he demonstrated the ruby laser, the Optical Society of America rearranged the program of its semi-annual meeting to schedule a major talk by him (Bromberg, 1991, p. 110), the only laser talk on the program. However, his achievement apparently was too practical for a Nobel Prize, which instead went to Townes, Basov and Prokhorov.

In 1987 Maiman received the Japan Prize for invention of the laser. Established by the Science and Technology Foundation of Japan, it has a monetary award similar to that of the Nobel, and is intended to reward more practical achievements (other recipients have included the people behind the "Green Revolution" in agriculture). Maiman has been inducted into the National Inventors Hall of Fame, has been elected to the National Academy of Science and the National Academy of Engineering, and is a fellow of both the American Physical Society and the Optical Society of America.

One continuing testimony to Maiman's achievement is the continued use and commercial availability of ruby lasers more than three decades after their discovery. The helical flashlamps used in the first ruby laser were long ago supplanted by linear lamps in elliptical cavities—an idea Maiman himself had anticipated. The vast majority of solid-state lasers today use other materials, primarily crystals and glass doped with neodymium. Yet ruby lasers still occupy a few specialized niches in the laser world of 1990.

References

Mario Bertolotti, *Masers and Lasers: An Historical Approach* (Adam Hilger, Bristol, UK, 1983).

Joan Lisa Bromberg, *The Laser in America: 1950–1970*, (MIT Press, Cambridge, Mass., 1991).

James Cavuoto, "Laser Pioneer Interview: Theodore H. Maiman," *Lasers & Applications 4* (5) 85–90 (May 1985).

Jeff Hecht, "Photo finish for the laser pioneers," *New Scientist*, 106, 43 (16 May 1985).

Laser Focus, "Xonics' acquisition of Hadron brings a strong new force to laser community," *Laser Focus* 28–32 (May 1977).

George F. Smith, "The Early Laser Years at Hughes Aircraft Company," *IEEE Journal of Quantum Electronics QE-20* (6) 577–584 (June 1984).

PETER SOROKIN

The Second Laser and the Dye Laser

Peter P. Sorokin was born in Boston, the son of a sociology professor at Harvard University. He studied physics as an undergraduate and graduate student at Harvard, receiving his PhD in 1958 for a dissertation on nuclear magnetic resonance. In 1957 he joined the research staff at the IBM Thomas J. Watson Research Center and in 1968 was named an IBM Fellow. He continues to work in laser science at IBM Research. In 1960, he and Mirek Stevenson demonstrated the solid-state uranium laser—the first new laser to operate following Theodore H. Maiman's demonstration of the ruby laser.

In the mid-1960s, Sorokin and John Lankard studied the optical properties of organic dyes, first developing the saturable absorber Q switch, then the dye laser. More recently he has worked in nonlinear optics and spectroscopy, developing techniques including four-wave mixing and time-resolved infrared spectroscopy, and pursuing a longstanding interest in the two-photon laser concept. A fellow of the American Physical Society and the Optical

Figure 7.1 Peter Sorokin (left) and Mirek Stevenson (right) adjust their uranium laser in 1960 (courtesy of IBM).

Society of America, he has been elected to the American Academy of Arts & Sciences and the National Academy of Sciences.

Jeff Hecht conducted this interview on September 27, 1984, at the IBM Thomas J. Watson Research Center, Yorktown Heights, N.Y. It was updated in 1990.

Q: How did you get involved in physics?

Sorokin: I owe that to Professor [Nicolaas] Bloembergen, my thesis advisor at Harvard. I had planned to go into theoretical solidstate physics. In my second year of graduate work, another student, Don Weinberg, and I signed up for a reading course on nuclear magnetic resonance given by Bloembergen, which we thought would be an easy course. At the time Bloembergen wasn't a very polished lecturer, and we let everything go over our heads, but we weren't too concerned. Then he announced that he wanted a term paper from each of us as proof that we took the course, and we wrote the papers and handed them in. We got unsatisfactory grades, so we went to Prof. Bloembergen, who said "These papers don't say anything about what I was teaching."

We couldn't get unsatisfactory grades, so I spent part of the summer trying to understand NMR, then wrote up a term paper which Bloembergen accepted. By that time, I felt I had invested so much time in the subject, which actually seemed interesting, that I might as well sign up and do a thesis with him. So did Don.

First Bloembergen assigned me a theoretical problem, and for a year I sat at a desk in the Gordon McKay Laboratory with a pad of paper. Finally I came back to him and said, "The divergent parts cancel, and all you have left are terms that are very hard to evaluate, but they're finite." And he looked at me and said, "Well, Peter, I think you'd better do experiments."

I was a little shaken, but I followed his advice. Then I was issued what was called a Pound box [Pound-Knight-Watkins spectrometer for NMR measurements] and told that a suitable thesis topic would be to measure the cesium chemical shifts in cesium halides. However, the cesium resonances appeared to have long relaxation times, which caused the NMR signals to saturate and disappear on resonance. It was very difficult to measure the chemical shifts with any accuracy. Another year went by, and I was beginning to get discouraged. I would come home to my parents' house late at night, often quite dispirited. My father, a professor of sociology at Harvard and an early riser, would start a tirade early in the morning, asking, "What sort of an ignoramus are you?" He was worried that I might have picked the wrong field. In his department, if a person had been in graduate school three years, the faculty met privately and said, "This person is probably hopeless. What do we do with him?" I had been in graduate school for four years, and I wasn't getting anywhere. I really thought I might quit.

At that point, a young scientist named Al Redfield came to Bloembergen's lab as a postdoc. He had just published a theory of saturation of the absorption and the dispersion of NMR, and he impressed all the graduate students with his ability to design electronic apparatus. He built his own version of an NMR spectrometer, based on the crossed-coil approach of Prof. Felix Bloch's group at Stanford. I began to sense that the cesium halides might be suitable for demonstrating aspects of his theory. Also, a pulsed version of his nuclear spin induction spectrometer might be very suitable for studying the easily saturated cesium resonances (via free induction decay). Redfield is now a professor at Brandeis University and has emerged as a towering figure in science. In the years since he was a postdoc in Professor Bloembergen's lab, he has played the major role in showing how to use NMR to elucidate in detail the structures of molecules with molecular weights as large as 12,000.

I dropped my attempt to measure these resonances with the Pound box, and built a double resonance, crossed-coil apparatus. One day, right after supper, I went back to the lab, turned on my newly built equipment, and observed an enormous signal. I repeated the experiment and discovered what caused the signal enhancement. I got really excited and instantly realized that I had a thesis. I showed this to Bloembergen the next day, and he, too, got all excited. From then on, he came every day to the lab to check my progress and advise me. In about half a year, I finished my thesis, and Bloembergen and I published a paper based on it. That experience set my course in science. The fact that I almost quit, but then found something just following my own hunches, gave me a lot of confidence, and I've retained that confidence all through my career.

Q: How did you get interested in the laser, or the "optical maser" as it was called in those days?

Sorokin: I was subsequently hired by IBM to work with Dr. William V. (Bill) Smith on microwave resonance in solids. When the famous paper on optical masers by [Charles] Townes and [Arthur L.] Schawlow appeared in the December, 1958 *Physical Review,* Smith suggested that we redirect our efforts to this new field. Along with Mirek Stevenson, who had obtained his PhD with Townes a couple of years before, and who had been hired by Smith's group at about the same time I was hired, we decided to get involved with this "optical maser" idea. That, incidentally, was Townes-Schawlow terminology. Gordon Gould hadn't coined the name "laser" yet, but he soon would. After he proposed the term "laser," Bell Labs refused to use it and in effect discouraged others from using the term, but it won out anyway.

We all went to the first quantum electronics conference [in mid-1959], held at the Shawanga Lodge in the [Catskill] mountains, right across the Hudson River. When we came back, we decided to drop immediately what we had been doing in order to focus on the possibility of finding a solidstate laser material. We were thinking of continuous optical pumping schemes, with lamp power levels on the order of watts. I was doubtful that mirrors with high enough reflectivities could actually be fabricated, so I conceived of using a polished square [of laser material].

The idea was that the light could make many passes through the medium, bouncing off the edges of the square in a low-loss manner by total internal reflection. If you then bevelled off just a tiny bit of a corner, that would give you a way to couple out light from this resonator. A low-gain system might thus be made to oscillate. Bill Smith and Rolf Landauer [also at IBM] thought it was an interesting idea, and Smith additionally pointed out that it could be made mode-selective. If the refractive index were just slightly greater than the square root of two, only those modes corresponding to waves exactly incident at 45° on the edges of the square would have low optical loss. Calcium fluoride has an index just slightly greater than the square root of two, so I said, "Aha! We'll try to find an ion we can pump in a laser scheme and see if it can be incorporated into a calcium-fluoride lattice." Rare earths looked good because of the protected $4f$ shell, so I went through all the journals to find out who had put rare earths in calcium fluoride. In the Russian literature, [P.P.] Feofilov had reported two very striking systems in calcium fluoride. One was trivalent uranium, very similar to a rare earth, with fluorescence at 2.5 micrometers. We figured that to have a low threshold, you should pump to a broad band, then drop to a metastable emitting level, then emit to a thermally unpopulated level.

Q: The classic four-level scheme?

Sorokin: Yes. We realized from Feofilov's published fluorescence spectrum that most of the uranium emission was on a transition

Figure 7.2 Sorokin firing a flashlamp-pumped dye laser built by Jack Lankard in 1968 (courtesy of IBM).

with a thermally unpopulated lower level. It also was apparent that $CaF_2:U^{3+}$ strongly absorbed in the visible and would thus be spectrally matched to a high pressure xenon arc lamp. The other system, studied by the same people, was divalent samarium. Its spectrum looked very promising also, provided the crystal was cooled to cryogenic temperatures.

Q: Did uranium require cooling?

Sorokin: Yes, to at least [liquid] nitrogen temperature to take full advantage of the four-level scheme. Samarium absolutely requires cooling to 20 K. To try these systems, we had to find somebody to grow the crystals under reducing conditions, because both uranium and samarium normally have higher valences.

Isomet [then] in New Jersey said they would make a run and dope calcium fluoride with samarium. We asked Walter Hargreaves at Optovac in North Brookfield, Massachusetts, to grow some calcium fluoride boules doped with uranium. He probably thought we were crazy, but he said he would do it if that's what we wanted, but we had to provide the uranium. So we sent him some rods, and got a phone call, "How do I cut pieces small enough to put in crucibles?" I said, "Use your native Yankee ingenuity!" Later, when I asked him how he solved the problem, he said, "I used an ax."

When we received the first shipments, I was thrilled because the uranium doped crystals were beautiful ruby red, and Isomet's samarium crystals were beautiful dark green. You know you've got the right valence when the crystals are strongly colored. Then we sent the crystals to Karl Lambrecht in Chicago to polish into rectangular shapes. It was just at this point that I heard on the radio that somebody from Hughes had announced he had an optical maser.

Q: Were you surprised that the ruby laser worked?

Sorokin: Yes. I don't think that anyone thought it would work that quickly. I believed Ted Maiman's results because he had done good work with microwave masers. But we were astounded at how he pumped it. He clearly must have had megawatts of light power, and we were thinking in terms of pump power of a few watts.

Stevenson called up Schawlow, whom he knew, and Schawlow told him about the photographer's flashlamp Maiman had used for an optical pump. Stevenson ordered one right away. We decided to forget about the polished square approach, because all this pump power was now suddenly available. Besides, we figured that our crystals would lase at a thousand times less pump power than ruby because of the four-level scheme. So we had Karl Lambrecht cut cylinders out of some of our other boules, and we had the ends silvered. We put the uranium crystal in a dewar equipped with an optical port, and set the flashlamp adjacent to this port. We figured that this somewhat inefficient optical coupling arrangement should still work, because the threshold for lasing of our crystal should be so much lower than that for ruby. The first time we tried the experiment, in November, 1960, it did work, and we saw the oscilloscope trace go off scale. Samarium worked for us the same way when we tried it a few weeks later, right at the beginning of 1961. These were the second and third lasers on record.

After succeeding at making uranium lase, we wrote an article for *Physical Review Letters*. Stevenson, being direct and aggressive, said, "We're not going to send it. We're going to drive down to Brookhaven and tell Sam Goudsmit [the editor] we want a decision before we leave." I said, "Mirek, you can't do that." "Nope, we're going to do that." So we got into a car and went to see Goudsmit. He was slightly confused about the differences between masers and lasers, and said he didn't want another "maser" paper.

Q: That was how Maiman's ruby-laser paper was rejected.

Sorokin: We were aware of that. Mirek talked Goudsmit into accepting it, and as we were leaving, Goudsmit said, "Next time, tell your people from IBM not to come down here with machine guns."

Q: How did you get involved with the dye laser?

Sorokin: Stevenson had been running a mutual fund on the side, and when top management told him he had to pick between IBM and the mutual fund, he left IBM. In 1964 I was working with John Lankard, whom I had hired as a technician in 1960. He has been in IBM Product Development for the past 10 years or so, where he has almost single-handedly driven a pioneering effort to put excimer lasers into computer production lines.

We started to try to develop a passive Q switch, which we thought we could do with an absorber that bleached at high light intensities. When I began looking for a substance with strong absorbance at the ruby wavelength, I quickly found that organic dyes had huge oscillator strengths and were highly absorbing in the visible. In particular, I found that phthalocyanines—complexes with metal ions at their centers—all absorbed near 694.3 nanometers. Chloro-aluminum phthalocyanine looked best, so we asked a colleague of ours, John Luzzi, to synthesize some. John was a very generous person; he didn't make us a gram, he made us a whole pound. That was to be important later, because it actually led to the discovery of the dye laser.

We placed these phthalocyanine solutions right in the cavity of our ruby laser. We fired the laser and, sure enough, instead of the usual train of spikes, out came a 20-nanosecond giant pulse. The thing worked so simply, so well, and it was so durable.

The spectral properties of the phthalocyanines were quite striking. They luminesced if you picked the right metal ion; thus, stimulated emission was one possibility. Another possibility was that one could get resonantly enhanced stimulated Raman emission, which actually wouldn't have been particularly noteworthy, but which seemed interesting to me at the time. Chloro-aluminum phthalocyanine was unusual in that it was quite soluble in ethyl alcohol, but its absorption peak in this solvent was shifted somewhat from 694.3 nm, so we had never tried this particular combination as a Q switch solution. I thought it might give the Raman effect, and at about 4:30 pm on Friday, Feb. 4, 1966, we zapped it with our big Korad ruby laser and took a spectrum of the scattered light with our trusty old spectrograph. The plate had a black smudge, so we knew we had something, but I had to leave.

I thought about it over the weekend, and Monday I told Jack, "We have to take another shot, only let's align some mirrors with the cell." We did this, then fired the ruby laser. Jack came back from developing the plate with a big grin on his face. There was one place in the plate that the emulsion was actually burnt. There was so much light we knew we had something. It was not stimulated Raman emission because it was right at the peak of the dye fluorescence. It was laser action in the dye.

Q: What did you do with the dye laser after that?

Sorokin: Initially we pumped the laser transversely. Bill Culver, who was then at IBM Federal Systems Division, suggested end pumping. It worked, and we got a beautiful beam. Then we decided to place an absorber in the cavity. It wasn't clear that we could make a Q switch, because there was basically no energy storage in the dye, but we wanted to see what would happen if we added a polymethine dye that absorbed at the wavelength of the phthalocyanine dye laser. Ernest Hammond, a summer faculty visitor in our lab, took the absorption spectrum of the polymethine dye we had selected. He told me that the dye also fluoresced very strongly.

Figure 7.3 Sorokin today (courtesy of IBM).

I asked if it absorbed at the ruby wavelength, and he said it did. So we said, "Let's try that. Maybe that works as a laser, too." We tried it and it worked, so we began to see that the effect was pretty general. We got all the dyes we could get from the chemical supply houses, even from the chemical stockroom here. We got our ruby going on the second harmonic, and just began pouring dyes into cells and pumping them to try to make them lase. Many of them did. I remember one afternoon we went down the aisle here at our lab asking our colleagues, "What color do you want?"

I really missed the boat completely on one of the most important things about dyes: Their ability to produce tunable, monochromatic light. The persons who discovered that, but never seem to get much credit for their efforts, were Bernard Soffer and B. B. McFarland at Hughes. In a brilliant experiment, with a diffraction grating set up as one of the mirrors, they showed that, not only did the dye laser spectrum collapse to a narrow line, but also that you could continuously tune it by rotating the grating. We just hadn't thought of it.

The next year, 1967, we published a long paper on our cumulative dye laser research up to that time. It was published in the *IBM Journal of Research & Development,* which was where we published the first dye laser paper. I liked working with the editor, Hunt Gwynne, who was interested enough in our work to insist that it be written up clearly and carefully. He read our article and said, "How about a conclusion? Any long article likes a nice conclusion." I thought and replied, "What about the possibility that the dyes could be flashlamp-pumped?" That was the trigger for our next experiments.

We had heard that some people at Hughes had tried pumping dyes with big flashlamps and that it hadn't worked. D.L. Stockman at General Electric, who had proposed flashlamp pumping of dye lasers even before our ruby pumped results had been achieved, had calculated that it should be possible. He built a complicated, fairly fast flashlamp that would have worked on most dyes, but not for the one he picked, perylene. He published a paper on his negative results and just left it there. I had learned about triplet-triplet absorption while preparing the conclusion for our long dye laser paper, and figured that a flashlamp with something like a microsecond duration could excite most dyes and get them lasing before the triplet state filled and killed the laser gain.

Jack Lankard and I saw a picture in *Physics Today* of a coaxial, disk-like capacitor made by Tobe Deutschmann in Massachusetts, and thought that if we combined that with a coaxial lamp, we should be able to make a very fast discharge. We bought two capacitors, a small one we called Baby Bear, and another one we called Papa Bear. We assembled Baby Bear in the coaxial geometry and saw a discharge and some fluorescence, but no sign of a laser. We figured that might happen, so we decided to try Papa Bear. We were slowly charging the capacitor up to its rated voltage of 20 kilovolts. (The lamp was designed to self-flash when the applied voltage exceeded the breakdown voltage of the air inside the coaxial tube.) The laser was aimed at the wall or something, and at the last minute, as we were charging it up, I moved over to the other side to get a better view. At exactly the moment I moved, the lamp self-flashed. I turned around and Lankard said, "It worked. I saw the beam on the back of your shirt." The flashlamp-pumped dye laser indeed worked fine, and again we went down the aisle and asked what colors did people want to see, and we were busy demonstrating the new laser for everyone all that afternoon.

Q: I gather you didn't have to worry about top management coming around and saying, "You can't do this"?

Sorokin: That's right. In fact, after the discovery of the dye laser they made me an IBM Fellow, which has really given me complete independence. I'm here in Jim Wynne's group, but I can do whatever project I feel like doing. I'm lucky in that regard.

Q: What are you doing now?

Sorokin: In the early 1970's, we did a lot of four-wave mixing in metal vapors, both to generate tunable infrared and to generate tunable VUV. We also took an idea of Bob Byer's and generated 16-μm radiation in parahydrogen in a mixing scheme that's very close to what Los Alamos finally adopted [for molecular laser isotope enrichment in uranium], but then the isotope separation project went to Livermore instead. What we've studied recently is most applicable to laser induced chemistry. We take molecular spectra of reactions initiated by laser pulses. Everyone knows about CARS, coherent anti-Stokes Raman spectroscopy, as a way to take transient Raman spectra. What we've developed here is a technique called TRISP, standing for Time-Resolved Infrared Spectral Photography. This is a way of taking time-resolved infrared spectra with the same time resolution you would get in CARS, using 10-nanosecond lasers thus far.

We can apply it to the study of fast chemical reactions. We've done work in initiating thermal explosions of methyl isocyanide with a carbon-dioxide laser and watched the isomerization to methyl cyanide develop in time. We studied laser-initiated gas-phase explosions of hydrozoic acid, again initiated with a CO_2 laser. We've studied the photochemistry of chlorine dioxide. TRISP is a real workable technique, and we used it to study transient excited-state absorption of a molecule (DABCO) we were considering as a two-photon laser candidate. We're pretty proud of TRISP, but so far there are no other takers, although Dave Moore at Los Alamos is apparently looking into it. Maybe that's because it's a little complicated. It's somewhat harder to set up than CARS, and you need a heat-pipe oven for the metal vapor, but we're fairly comfortable with it.

It would be very interesting to look at a molecule hit by a pulse while it's still in the process of photodissociating. Chemists talk a lot about mechanisms, but it's mostly in the form of hypotheses, deduced by observing what products are finally formed. If you could capture the dissociation process spectrally, it would be very exciting, but you do need subpicosecond time resolution.

We tried to develop TRISP into a subpicosecond technique, but we found it exceedingly difficult to generate broadband femtosecond infrared continua (in other than a very limited spectral range) with the same technique of stimulated electronic Raman scattering that had worked in the nanosecond regime. However, we did find that broadband femtosecond ultraviolet continua could very easily be produced by focusing intense femtosecond ultraviolet pulses into high-pressure rare gases. Accordingly, we shifted our interest for transient absorption spectroscopy experiments to the ultraviolet.

With intense ultraviolet pump pulses provided by femtosecond pulse amplification in excimers, and with continuum probe pulses generated by the method I mentioned earlier, we now have the capability to perform transient ultraviolet absorption spectroscopy experiments on a femtosecond time scale. For the past two or three years, my present colleagues, James H. Glowina and James A. Misewich, and I have been applying this apparatus to study gas-phase photodissociation of a number of systems, including thallium halides, diatomic bismuth (Bi_2), sodium iodide, and iodine cyanide (ICN). We excite the molecules with the pump pulses, then we use the femtosecond ultraviolet continuum pulses to take spectral "snapshots" of the molecules while they are in the process of falling apart. The transient spectra so far have shown several surprising features, which have required us to change our ideas considerably about the photodissociative process.

I'm always on the lookout for a two-photon laser, which would emit pairs of photons. We looked at one promising candidate (DABCO). After we did our homework, we found that the cross-section is a factor of five too small, which is, however, getting much closer than we've been before. I still regard the two-photon laser as a potentially interesting and realizable device. Within this year we will be trying another such system, this one based upon organic dyes.

Q: What special advantages would a two-photon laser have?

Sorokin: A two-photon laser is so hard to build that I don't think it would occur naturally, as for instance does the maser. Somebody out there who has a two-photon laser and wants to let the universe know could send its beam out into space. Millions of light years away, when someone detects it, the photons will always come in pairs, showing that this is unusual light. There are some research possibilities; I think people are interested in the coherence properties. But I cannot honestly imagine another particular advantage of the two-photon laser.

Q: What do you think of dye-laser technology today?

Sorokin: I never expected it to be this good. Right at the beginning, its continuous tunability and monochromaticity—Soffer and McFarland's result—was a tremendous surprise; I think that affected me more than any subsequent dye laser development. Then there was [Ben] Snavely's cw laser, and all the work of [Charles] Shank and [Erich] Ippen in modelocking the dye laser. I'm in great awe of the recent technique invented by Dan Grischkowsky and his co-workers to compress pulses from commercially available dye lasers down to a few tens of femtoseconds in a totally passive way with the use of optical fibers. I understand how it works, but I never would have expected it. The flashlamp-pumped dye laser hasn't been that much improved; it's still kind of a wild beast. But the demonstrated usefulness of dye lasers in spectroscopic experiments is just astounding.

Q: If you had it to do all over again, would you still investigate lasers?

Sorokin: Oh yes. The field has been very kind to me, IBM's been very good to me, and I certainly feel that I've been very lucky. The real advantage I had was being right there when the field was starting. My own abilities are such that I'm better at doing the first-order investigation than the subsequent careful follow-up. Coming into science in the late 1950s, I was very fortunate to come into a field about to explode, where I could make contributions, and I've been well-rewarded, too. So the answer is a resounding yes.

An earlier version was published in *Lasers and Optronics* ®
(formerly Lasers and Applications) a Gordon Publications, Inc. publication.

ALI JAVAN

The Helium-Neon Laser

Ali Javan, a native of Iran, came to the United States to attend Columbia University, where he received a PhD in 1954 for microwave spectroscopy work performed under Charles H. Townes. After four more years at Columbia as a research associate and instructor, he joined Bell Telephone Laboratories in 1958. At Bell Labs he continued theoretical and experimental studies on gas discharges and laser physics.

This research culminated in the December, 1960 demonstration of the first gas laser, a helium-neon type emitting at 1.15 micrometers, by Javan, William R. Bennett Jr., and Donald R. Herriott. Soon after, Javan joined the faculty of the Massachusetts Institute of Technology, where he now is Francis Wright Davis Professor of Physics. He founded and headed the MIT Optical and Infrared Laser Laboratory. He is also the founder and chairman of Laser Science Inc. in Cambridge, Mass.

Javan received the Franklin Institute's Stuart Ballantine Medal in 1962 and the Optical Society of America's Frederick Ives Medal in 1975; he is a fellow of the National Academy of Sciences,

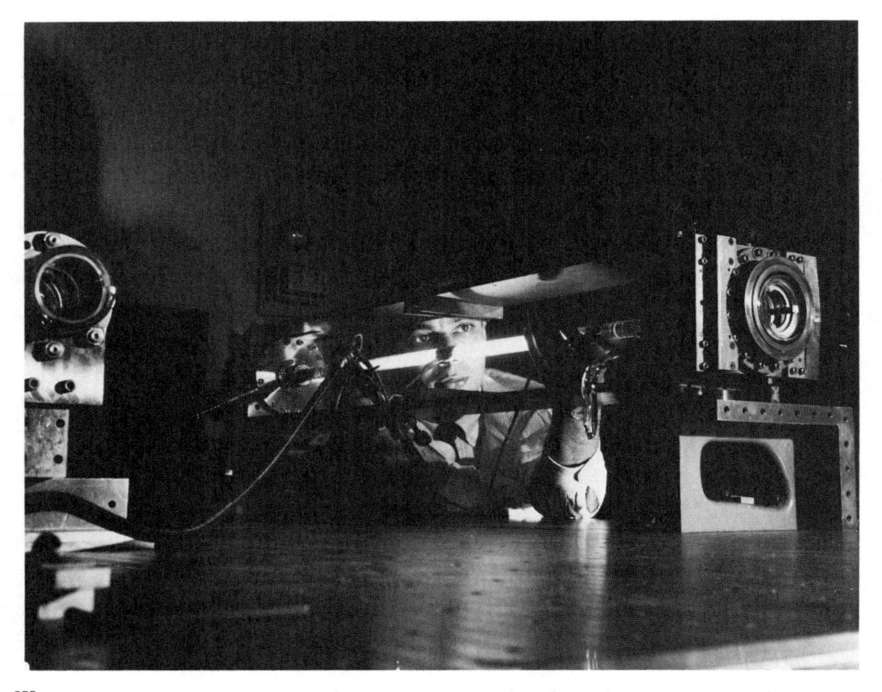

Figure 8.1 Ali Javan demonstrates the first helium-neon laser at AT&T Bell Laboratories (courtesy of Ali Javan).

the American Academy of Arts and Sciences, the American Physical Society, and the OSA.

Jeff Hecht conducted the initial interview on February 28, 1985, in Javan's office at Laser Science Inc., and supplementary interviews at Javan's home on September 18, 1990 and at MIT on April 5, 1991.

Q: How did you first get involved with lasers?

Javan: Science has been in my blood since I was a little kid, and I was always fascinated by light and radiative processes, so I got into physics. I did my PhD thesis on microwave spectroscopy at Columbia University under Charlie Townes. I was not working on the microwave maser then, but I could see the whole field evolving. Those were exciting days at Columbia, with Charlie, [I.I.] Rabi, and Willis Lamb discovering masers, the electron anomalous g-factor, and the Lamb shift.

After my thesis, I became more involved with masers when Charlie was on a year's sabbatical leave in Paris. My interest led to my discovery of the idea of a three level maser, in fact long before a group from Bell Labs published the first experimental work on the subject; actually, about a year before. I delayed publication because I became fascinated by a coherent, two-quantum effect, similar to a Raman process, which takes place in the three level maser. That, in turn, led me to the discovery that a Stokes-shifted Raman transition can produce amplification without requiring a population inversion. I published my three level maser theory in 1957 and the Raman work in a French journal in 1958.

Charlie Townes was encouraging me then to think about submillimeter and microwaves because we had resonators at these long wavelengths for feedback. Although I was aware of the possibility of achieving gain at short wavelengths, I did not know how to make a laser because I was not aware that a two-mirror Fabry-Perot optical resonator could be used to introduce the feedback.

About this time, Sid Millman arranged an interview for a job at Bell Labs and I talked with Art Schawlow there; he told me about the idea he and Charlie were exploring to make an optical resonator. I have always had a close relationship with Charlie, and I am really very grateful he did not talk to me about his ideas on lasers because that way our ideas did not get mixed. I was able to develop my own thinking.

After I learned that a Fabry-Perot might induce optical feedback, I rushed back to Columbia and started to calculate the same afternoon. I never really read the Townes-Schawlow paper [in *Physical Review*] because, by the time their paper came out, I was already deeply involved in my gas laser work. Even before I started at Bell Labs, I realized that, for a gas laser, you needed to find a pumping mechanism better than optical pumping. I should underline that I always looked to gaseous media. I am a gas-laser person; I don't do solids. I prefer the simple interactions of single atoms or single molecules with a radiation field. The Townes-Schawlow paper primarily related to cesium and some other species pumped by an incoherent light source. I started looking into other possibilities, and that led to my discovery of gas discharges as media where non-equilibrium conditions could lead to the presence of inverted populations.

I went to Bell Labs in 1958. There a dilemma evolved. RCA had previously inspected my notebooks on the three level maser and had established that my dates preceeded the Bell dates. They had paid me $1,000 for patent rights and were contesting Bell Labs' application. For the first six months or so working at Bell, I was dealing with RCA *and* Bell patent attorneys; it was very uncomfortable. Luckily, RCA did some marketing studies and concluded that the maser amplifier was not commercially viable. Thank God for that. We agreed to drop it and let Bell have it.

When I got to Bell, they had just dismantled their gas discharge department. My boss, Ted Geballe, told me to go buy a Varian magnet. I told him I didn't know what to do with it, but he told me to buy one anyway because sooner or later I would be needing it to study solids. So I ordered one and didn't even use it once. I published my initial proposal for producing population inversions in gas discharges in *Physics Review Letters* in 1959, and went after discharges on my own.

Q: A lot of the other early lasers, both gas and solid-state, have just faded away, but helium-neon is still widely used. How did you manage to make such a successful choice so early?

Javan: It was no accident that the first gas laser is the largest-selling gas laser, with almost half a million a year produced now, worth maybe $50 to $60 million a year just as components. I made a careful selection of a system with the promise of being the best medium for the first laser, and that best medium has stayed with us until now.

Helium-neon was one of the cleanest systems I could find. It also was a medium where I could show there was gain without first having to make the laser. Even after I had convinced myself that helium-neon was the best gas medium, there were a lot of non-believers telling me that gas discharges were too chaotic. They said there were a lot of uncertainties, and I had nothing I could control.

Q: How did you go about proving them wrong and building the laser?

Javan: Nowadays if you have an idea of a gas system that might lase, you can align two mirrors two or three meters apart, put in the gas, and see what comes out. I couldn't do that because I didn't have a laser to align the two mirrors.

I decided to first establish that I had gain in the tube, then to try to align the mirrors by trial and error. That is how it worked. Without all my preliminary work to preset the HeNe discharge at a known gain, it would have been impossible to make it work. You could not have varied that many parameters and kept wiggling the mirrors without knowing that you have gain.

Q: You started out working alone. When did others join you?

Javan: I always enjoyed working with others; otherwise this business would be too lonely. After I had done a lot of the early work, I persuaded Bill Bennett to come to Bell Labs for a year or two to join me. He had been a friend at Columbia, and was on the faculty at Yale University. He was very helpful, and I remember working late nights with him for over a year.

I also worked with Ed Ballik, who came to me as a technician, but was really a high-level person who made basic contributions. You don't hear his name much, but he was a very important part of the team. Later he got a degree at Oxford University, and now he's teaching in Canada.

The initial impetus to get Don Herriott came from Al Clogston and some other people at Bell, because he had been in optics before. It was essentially the four of us, with the support of some other people at Bell Labs.

Q: How did you get from observing gain to having a laser?

Javan: I had the gain six or eight months before I had the laser working. I could see the gain, but I had non-believers. I had measured the excitation transfer from helium to neon, but people wouldn't believe that I had, in fact, measured the right cross-section. A year later, when I had the final measurement, that initial transfer cross-section turned out to be exactly right. But the non-believers said the gain I saw was really some nonlinearity in my detection system.

That really got me going. "By God," I said, "I'm going to make the damn thing work." I decided that the gain was going to be there. A lot more went into it, but the system finally worked on December 12, 1960; I remember it was 4:20 in the afternoon and it was snowing.

One interesting thing happened just after it worked. About six months earlier Ed Ballik had brought in a bottle of wine that was a hundred years old. We kept it to open after the laser worked. A few days later, I called the head of Bell Labs and invited him to come have this hundred-year-old wine. He said he would be very glad to come, then said, "Oh, oh, Ali. We have a problem!" This was two or three in the afternoon. He wouldn't tell me what the problem was, but said he would come at 5:30. Later that afternoon, a memo was circulated through the lab. It turned out that some months earlier they had disallowed liquor on the premises. The new memo stated no liquor was permitted unless it was over 100 years old. After that, he came.

After it worked, my friends in the Bell Labs administration finally told me that other people in the administration had called my efforts a wild-goose chase. In fact, they had been talking of cutting off my crazy idea in three or four months. That made me shiver.

Q: I have heard of other people having such problems. Hughes' management had told Ted Maiman to stop working on the ruby laser, but he went ahead anyhow. Do you think you might have gotten into that situation?

Javan: I do not think so. Even if it hadn't worked when it did, I am sure I could have convinced people at Bell to keep it going. It would be too dramatic to say that I would have been cut off. My friends in the administration told me with pride that they defended my project. Bell labs had made a commitment; they told me the whole thing cost over $2 million. I am not so sure about the figure; maybe it included the salaries of all the top brass of Bell Labs. I was spending a lot of money, though, and if it would have gone another year and the damn thing wouldn't have worked, then there would have been real problems.

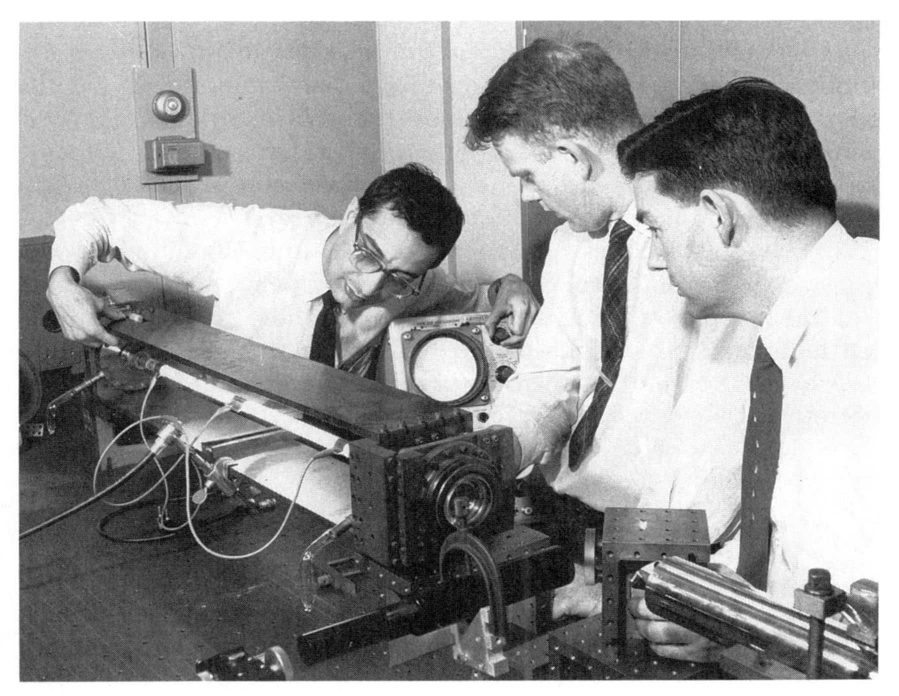

Figure 8.2 Ali Javan, William R. Bennett Jr., and Donald R. Herriott (left to right) adjust the first helium-neon laser in 1961 at Bell Labs (courtesy of AT&T Bell Laboratories).

Q: What do you consider most important about the helium-neon laser?

Javan: We were looking for coherent light, and that was what was special about the helium-neon laser. The ruby laser was not coherent; it emitted a pulse of light, which was a hash. The light from the helium-neon laser was coherent, and it was the first really coherent laser.

Q: Your first helium-neon laser operated at 1.15 micrometers in the near-infrared. Why not the 632.8-nanometer red line?

Javan: I was aware of other transitions, but I picked the 1.15-μm line because of external restrictions. I could detect this wavelength with a photomultiplier, and the gain at the shorter wavelengths was smaller according to my calculations. My study showed that

1.15μm was where you had the best chance to get the highest gain and the best experimental conditions to show the gain and optimize it. Also, the branching spectrum from the neon fine-structure was better resolved than on other transitions. So we could do the spectroscopy to show we could get the gain above the 7% or 8% needed for the laser with 99% reflective mirrors.

If we had set up the systems for the red line, the helium-neon laser would not have worked then. To get high gain on the red line, you have to have tubing with such a small bore that you cannot see through it to align the mirrors with a conventional autocollimator. Remember, in those days we were working only with flat mirors. Now we use curved mirrors, which are easier, and we have other lasers we can use to line up the mirrors.

Q: Was it fairly straightforward to take the step from the infrared line to the red?

Javan: Everybody who made a contribution beyond the first set of laser transitions played an important role. Still, the red transition followed naturally and logically. The laser could be aligned already on the 1.15 μm line, and there was a lot known about the whole process. Later hundreds of transitions were discovered in HeNe lasers. One, at 3.39 μm, has such high gain it would even oscillate with the reflection from the back of your fingernail. All hell broke loose, but by then MIT had offered me a job. I really just passed through Bell Labs and dropped the helium-neon laser when I left. Bell Labs had done a very wise thing and encouraged other people to get into it within a month after my laser worked. The gas discharge department was re-established overnight.

Q: The neon fluorescent tube was known long before anyone thought of the laser, yet it didn't seem to lead anywhere. Do you have any ideas why?

Javan: If you look back on the evolution of quantum mechanics and spectroscopy, you find that gas lasers, maybe a helium-neon laser, should have been discovered in the early 1930s. It shows that, even if the time is right, something may be totally missed. The German spectroscopists were studying gas discharge media and would plot the population of excited states versus discharge current. In the early 30's they studied neon extensively. A little extrapolation of their work would have given rise to an inverted population. By that time physicists knew the expression for negative dispersion, and there was a German paper with the subtitle "negative dispersion in neon." But at that time, when people experimentally saw a population with nonthermal distribution, they would immediately try to bring it back to thermal equilibrium. That meant no gain and no lasers.

Why was that? They were fascinated with thermal equilibrium because that led to Planck's discovery of black-body radiation law and the Einstein A-coefficient, the cornerstones of the discovery of quantum mechanics. This fascination with systems in thermal equilibrium inhibited workers from thinking you could have non-thermal distribution, which could lead to gain. In fact, more of our universe is in non-thermal states than is in thermal states. Now at that point, lasers would have been possible only in gases because solid lasers would have required developments in crystal growing that only came after World War II.

There is also the question of feedback, but if you look in old optics books, you will find Schroedinger's interferometer, an experiment from the early 1930s. He took a fluorescent crystal, polished its two ends, made them parallel and reflecting, and illuminated it with a flash of light. He was looking for dipolar and quadrupolar radiation inside the resonant cavity. Fabry-Perot cavities were known from the days of Mr. Fabry and Mr. Perot. So gas lasers could have been discovered in the early 1930s, but by the mid-thirties people had presumably discovered everything they wanted to know about gaseous media and had moved to something else.

Suppose that gas lasers, including carbon-dioxide lasers, had been discovered before World War II. Then laser radar, not microwave radar, would have been the name of the game in World War II; microwaves would have come later. We would have now been working on microwave sources and radar. But because people missed the chance in the 1930s, we had to wait for Charlie and the maser, and then we were into lasers.

Q: What took you from Bell Labs to MIT?

Javan: I always wanted an academic career. MIT made me an offer. Charlie Townes had just joined MIT, and I'm sure he had a lot to do with it. I set up my research lab, which I built up to a very large operation. For a year after I got to MIT, I had a joint appointment with Bell Labs and kept a lab there too. The distance from Murray Hill to Boston made me drop that very nice relationship later.

Q: What sort of things did you do at your MIT lab?

Javan: I decided that I wanted to understand laser processes and do things with the laser. I have developed a whole lot of technology over the years. I did the original work in high-resolution laser spectroscopy. My original work on three level masers had turned out to be very basic in studying the interaction of an optical field with a pair of atomic resonances and branching to other resonances. The only new thing in optics is the presence of Doppler broadening in gases, which turns out to be quite a bit different than inhomogeneous broadening in solids. Willis Lamb's paper on the Lamb-dip was a very important contribution and revealed new features. In the 1960s, I introduced a number of ways to eliminate the Doppler broadening in atomic spectra and to obtain very high-Q resonances for very high-resolution spectroscopy. Later in the 1970s, I extended these to molecules in the IR. Now, with good dye lasers, a lot is being done in Doppler-free spectroscopy.

I take great pride in having originated the technology for extending microwave electronics into optics. This led to my high-speed diodes with a response time as short as 10^{-14} or 10^{-15} second. My work in this area led to the absolute measurement of light frequencies, in which the frequency of light is compared with a microwave clock. Now the National Bureau of Standards and others have followed my MIT work in their speed of light measurements. John Hall at NBS in Boulder has done a lot of beautiful work with considerable impact. My absolute wavelength measurement of Doppler-free lines in a CO_2 absorbing gas took me over eight years to do with my very talented students and colleagues.

I have unfinished business in the field I call "optical electronics," extending microwave electronics technology to the optical frequencies. I can envision oscillators and amplifiers at optical frequencies that would operate on the same principle as microwave oscillators. They would be classical devices in the same way that microwave or radio frequency oscillators are classical devices. These kind of elements wouldn't compete with lasers. They could serve as components for high-speed computer memories, in holographic imaging in real time, and in applications you couldn't even predict now. Such work requires access to special microelectronics which is now available. I have full intentions to follow this work at MIT.

Q: What are you doing now at MIT?

Javan: I have never liked to work in research areas that are crowded with a lot of people doing a lot of good and bad work. So much of my work over the years has evolved into big fields. For this and other reasons, I am now in the process of changing the direction of my MIT research. In fact, I've given myself a little time to establish the next research phase.

My plans are to get into studies of the upper atmosphere with lasers. This is already a developing field, but I plan to do it in other ways—to extend quantum electronics into studies of upper-atmospheric phenomena, not just linear-type laser atmospheric remote sensing. I intend to take advantage of nonlinear interactions between laser radiation and atmospheric species in a variety of ways.

Q: Is that what you're trying to do with Laser Science, the company you formed in Cambridge?

Javan: I did the planning for the company when I was on sabbatical leave a few years ago. My relation to LSI is, naturally, outside of my MIT activities and duties. The company is primarily an engineering operation based on the technology I have developed over the years. LSI is now introducing a whole new kind of laser technology that I know will have a lot of impact.

Most of the commercial gas lasers you can buy now were not just discovered, but were also perfected in the engineering sense, in the 1960s. A lot of things happened in the 1970s and 1980s, but where are they? What happened at the end of the 1960s is that the attention of the Department of Defense—which has been the prime force in pushing the engineering evolution of gas lasers—got shifted to high-energy lasers, mainly the CO_2 and the HF/DF lasers, which can be used to produce a lot of laser energy. You can be proud of the achievement in very high-energy gas lasers, but these very high energy lasers are not viable for commercial use. For that, you need lasers at medium or low energies.

Another reason that DoD shifted away from medium and low energy gas lasers in the 1970s was that neodymium-YAG was there. Now we find that YAG has great limitations for broad-based applications. Gas lasers can go deep into the ultraviolet and deep into the infrared; there are great varieties of them, and you can have precise control over the wavelength, even at a moderate pulse-energy. The medium and low energy range is where the action is, but nobody in the country has given it attention with respect to gas lasers. That's where Laser Science comes in—to develop medium and low energy gas laser technology for a variety of very large-scale DoD and commercial applications, where the market is now in the developmental stage.

When I mention low energy, I think of pulsed lasers producing 10 to 100 microjoules per pulse. If the pulses are in the nanosecond range, you can get 50 or 60 kilowatts peak power. You can do every nonlinear thing under the heavens with that much peak power. In particular, if you focus that energy to very small areas, you can get more than gigawatts power flux.

When I say medium energy, I mean a CO_2 laser, for instance, at energies of several joules and higher, with the possibility of obtaining very accurate and pure frequencies for coherent Doppler lidar and remote sensing. For that, the extraction of laser energy from an energetic medium at highly-controlled frequencies is the name of the game, and my company has the foremost expertise in this area and entirely new ways of doing it.

Q: How do you feel about laser technology today? Has it brought any surprises?

Javan: Let's limit ourselves to gas lasers. Just a few months after the helium-neon laser, a doctor came to see me and started telling me about medical applications. That came as a kind of remote possibility and a surprise to me then. I just listened to him with great respect. The idea of focusing light and looking at photochemical reactions for medical use, and things like this recent work of

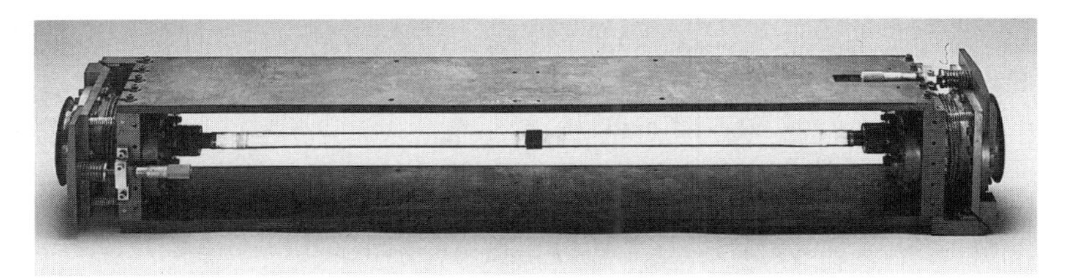

Figure 8.3 The first helium-neon laser, which Javan, Bennett and Herriott demonstrated in December, 1960 (courtesy of Ali Javan).

cleaning arteries, they're just beautiful. If anybody had said he was working on a laser to make an impact on surgery, you would have said he was crazy. So in that sense there were plenty of surprises.

But then, surprise really is not the right word, because things happened gradually. I really had no idea of things like laser fusion or high-energy laser weapons in those early days; none of us did. But I was not really surprised because they evolved little by little, and I was with them as they evolved.

Q: Where do you think gas lasers might go from here?

Javan: Before we think about new things, we have got to make the things we've already done in the laboratory really happen. To keep looking at the next thing and the next process is open-ended, though x-ray lasers and the like have to be done for the future. I think the technology of the '70s will happen and will have a large impact on industry and our very existence. I think the field is open for gas laser technology at medium and low powers. If you want to stick to a few milliwatts continuouswave at a fixed wavelength, it's forever helium-neon. But the new technology of making pulse intensities in the multi-kilowatt and higher ranges at nanosecond pulses available for use by the non-laser physicists, the biologists, and the chemists, and making them low-cost like a flashlamp, opens up large-scale commercial possibilities.

There are more of these things to come because the gas medium offers wavelength flexibility and the possibility of tuning. Laser radar has applications for peak power in the megawatt range extracted at a controlled frequency. The challenge is to extract that energy at a highly pure and stable frequency. You don't need a new laser—the CO_2 laser is a beautiful one. The challenge before us is to make what we need happen.

In terms of the future and beyond, making an x-ray laser is fine, but we should also look for other ways of making coherent x-rays, other than a laser process which requires dumping a lot of energy into the medium. And I'm not talking about harmonic generation. There are other ways of extracting coherent continuouswave energy that don't involve inverted population and gain or a parametric effect. I have some ideas on that line. The future is open.

[Additional comments, from interviews conducted September 18, 1990, at Javan's home in Cambridge, Mass., and April 5, 1991, at Javan's office at the Massachusetts Institute of Technology.]

Q: You're still involved with lasers after 30 years. What keeps you going?

Javan: It is hard to believe that it is 30 years since the early days of lasers. I teach graduate students at MIT who were not even born then. The laser field is still going strong; I have unfinished business in research that keeps me going. Most fields of research reach their peaks in five or six years after the initial discoveries. The laser field, although it has been around for thirty years, is still quite a bit like that; it goes through stages. With a new discovery, a new technological or research area begins to flourish; it goes on for five or six years, and in some cases a little bit longer, until it reaches its peak and then evolves into something else. Another discovery comes by, and the cycle repeats.

The promise of the future in the laser field is not the discovery of new lasers. The exciting possibilities are to make the existing lasers behave in new ways, and then use them in new ways, in a new technology area or new research. We can extract these days, for instance, a pure frequency from a diode laser, not as pure as can be extracted from a gas laser, but pure enough for an application such as coherent fiber-optic communications, an-up-and-coming field in long-distance communication using frequency multiplexing. To give another example, laser-induced fluorescence has been around since the early 1970s; the first experiment with it was a physics experiment at my MIT laboratory, though its primary use now is in biomedical applications, also an up-and-coming field. This is characteristic of lasers; old things come to the surface and are used in new ways. It will go on and on this way, clear into the twenty-first century. Light is the way we communicate with atoms and molecules; there is an infinite variety of ways to make use of it.

Q: Would you care to comment on military funding of laser-related R&D?

Javan: Military uses of lasers moved at a tremendous pace in the 1960s and 1970s because Washington spent the money to make it happen. A great deal of our progress today is due to that expenditure, including progress in areas beyond the military. In laser material processing, for instance, an industry of about $100 million in yearly sales, the lasers used today are the lasers developed in the 1970s with military R&D money for possible military uses, some of which did not pan out to be useful to the military. Most of the industrial lasers today are a result of earlier military R&D expenditures.

Our system is not geared to support rapid development of non-military applications requiring R&D in areas that do not have a foreseeable military use. We all want to take big steps, but the big steps require adequate commitment of funds. The pace of military expenditures in lasers has slowed down a great deal, even in areas of potential importance in defense. The industry now has to rely mainly on developments of the past decades, though the field is quite lively, and quite a lot is going on.

Q: What new medical applications do you expect?

Javan: I expect a great deal from what is going on now. With fiber-optics a surgeon can take laser light where he wants without making an incision, both for doing surgery and for diagnostics.

There is a great deal of excitement these days in medical diagnostics. Much of it is still in the research stage, but large-scale clinical uses are well in sight in this decade. Live-cell fluorescent studies, which do not kill or damage the living cell, are developing to be extremely important research tools in understanding the impact of various drugs on cell responses. Ordinary light sources have been used in the past for these fluorescent studies, but a big problem has been damage to the living cell by the light energy needed to induce the fluorescence. With a short-pulse laser of the right color, one can induce fluorescence with very little energy, without damaging the cell. There are breakthroughs these days in discovering new types of fluorescent probes for these types of studies. The probes are biological molecules that are selectively absorbed by the living cell and make it fluoresce in the parts of the cell where the molecule is absorbed. Calcium fluorescence is an important probe. The fluorescence is viewed through a microscope and imaged on a video screen.

Here is an example of an entirely non-military use of lasers that requires the development of lasers with special characteristics tailored for these types of biomedical applications. I don't know of a single government agency that would even want to hear a proposal to fund the development of a laser for a specific medical use. NIH [the National Institutes of Health] gives a good many grants for laser medical research, but the research must rely on commercially available lasers with whatever characteristics that happen to be available. An ultraviolet excimer laser, for instance, in a compact configuration with low pulse energy and short pulse duration, can have exciting uses in medical diagnostic research and ultimately in clinical applications. The excimer lasers that are available today are the spinoffs of past military R&D, and none of them are of much use in the kinds of applications I am talking about. The company I founded several years ago, Laser Science Inc., pioneered development of an ultraviolet nitrogen laser specially designed for these types of applications, with quite a bit of success. I think NIH should sponsor the development of lasers with specific characteristics needed for biomedical applications.

Q: I continue to be amazed at some research applications, like single-atom spectroscopy.

Javan: Yes, you can set a trap in a small vacuum chamber, an rf [radio-frequency] trap of a few cubic millimeters in volume, catch a single ionized atom, an ion, hold it for a long time, and look at it by shining on it laser light of the right color to make it fluoresce and scatter light. To hold the ion stationary, you shine on it another laser light of the right color and the right intensity, which exerts a force on the ionized atom that slows down its kinetic motion and holds it still. The experiment is a lot of fun, looking at a single ionized atom; it has intriguing possibilities in high-resolution spectroscopy.

Q: You mentioned earlier another intriguing concept, tiny optical oscillators based on classical principles. What would they do?

Javan: They are a part of my unfinished business to which I referred earlier. To give you some background, in the radio and microwave frequency regions, we have had classical electronic devices, oscillators and amplifiers and the like, in which the electric current flowing through the device elements determines their functions. We call these "classical" electronic devices because their descriptions do not require quantum mechanics; classical electricity and magnetism, based on Maxwell's equations, are more than adequate to formulate their functions, to design them, and to make them work. Lasers, on the other hand, are purely quantum mechanical devices; to describe them we need quantum mechanics, the Schrödinger equation, energy levels, stimulated emission, and the like. I have always said that if you let the Planck constant, h, go to zero, all the visible and infrared lasers would shut off, but the radio and television transmitters will go on merrily broadcasting their radio or microwave signals.

In the radio and microwave frequency regions we have both classical and quantum mechanical devices, rf and microwave oscillators and amplifiers, as well as maser oscillators and amplifiers, the maser being the microwave analog of the laser. I now ask the question: if in the radio and microwave frequency regions we have both classical and quantum mechanical devices, then how come in the infrared and visible regions, in optics, we have only quantum mechanical devices? That is, we have lasers, but not the classical types of optical-frequency oscillators, amplifiers, down-converters, upconverters, and the like.

Having said that, I must point out that two conditions must be satisfied in classical devices where the electric currents flowing through the device elements determine the device characteristics at a radio or microwave frequency. One relates to the speed of the device elements, their time constants, and the other to their size and dimensions. The speed must be faster than the operating frequency, and the size must be less than a wavelength. If the speed is not fast enough, the electric current will not follow at the frequency at which it is designed to operate. If the size is larger than one wavelength, then there will be phase cancellation, unless we consider distributed circuits, but even there each element in a distributed circuit will have to have a dimension less than a wavelength.

This says that for a classical electronic device operating at an optical frequency, we need high-speed elements with response times faster than the optical frequency, and with dimensions less than an optical wavelength. Both conditions are needed.

This takes me back to the early 1970s, where in a series of experiments at MIT, I introduced into the optical region a biterminal element, a diode, with a nonlinear current-voltage characteristic, an intrinsic response time faster than the optical frequency, and dimensions less than an optical wavelength, in fact in the submicrometer region. The element, by now quite well known, consisted of a metal-insulator-metal (MIM) tunnelling junction. In the initial version, the one most commonly used today, the junction was made by a point contact between two metals across a thin oxide barrier, with the contact area in the submicrometer region. Later on I was able to show that it is possible to form the high-speed junction using thin-film microelectronics on a substrate. In both versions light is coupled to the junction by an optical frequency antenna, a dipole, integrated with the junction.

This was the very first example of a classical device at infrared and optical frequencies, and the very first time an optical frequency antenna was used to pick up a light wave and couple it to a device element. This element has been the subject of a number of physics PhD theses at my MIT laboratory. I did this work to extend microwave frequency measurement technology into the optical region, by mixing in the junction the frequencies of two lasers differing by several octaves, for example, a far-infrared laser with an infrared laser, or a microwave frequency with a far-infrared laser. This was a breakthrough at the time. It is the key technology in optical clocks, a currently developing field of considerable importance in both science and technology, which "times" the period

Figure 8.4 Ali Javan today (courtesy of Ali Javan).

of an optical cycle using a cesium atomic clock operating in the X [frequency] band.

Q: You said you have unfinished business in this area. What is that?

Javan: No one has yet tapped the possibilities of forming a high-speed element with submicrometer dimensions on a substrate with thin-film microelectronics, which would respond to optical frequencies as a classical element. I recently established a research project at MIT to do that.

I have been aware of these possibilities for quite some time, have done quite a lot of background work, and now would like to apply the touches that it takes to give it a head start. I have an ongoing experiment at MIT, from which I hope to obtain results soon, using thin-film microelectronics to form new types of high-speed elements with thousands of times larger nonlinear current-voltage response than the earlier MIM elements that came out of my laboratory.

With thin-film microelectronics it should be possible to integrate the high-speed elements with a tiny tank circuit resonating at an optical frequency, an LC [inductance-capacitance] element resonating at 10^{13} to 10^{14} hertz. The inductance in the tank circuit can be a loop 3 or 4 micrometers in diameter or less, integrated with a submicron-size capacitor. With state-of-the-art x-ray or electron lithography, one can form a structure with several hundred ångstrom resolution. Other lumped circuit elements can be introduced into the infrared and the red region of the spectrum. Distributed circuits can be formed on a substrate, with phased-array characteristics at optical wavelengths, using a large number of micron-size classical elements coherently for holographic signal processing. Classical oscillators and amplifiers are possible at optical frequencies. The types of oscillators and amplifiers I am talking about will have new uses in which the alternating currents at optical frequencies flowing through the high-speed elements can be manipulated in new ways. Multiterminal elements of submicron dimensions responding to the flow of alternating currents at optical frequencies could be possible later. This is the stuff of the future; I could go on.

Q: What about measuring time and gravity waves?

Javan: An atomic clock makes use of an atomic resonance to define the precise frequency at which the clock operates. An optical clock makes use of an optical resonance, atomic or molecular, to define the clock frequency. The primary frequency standard these days is the cesium atomic clock operating at an X-band frequency in the microwave region. Since the early days of Doppler-free laser spectroscopy in the 1960s and early 1970s, we have had high-Q Doppler-free resonances at well-defined frequencies in the infrared and visible with characteristics well suited for highly accurate optical clocks. However, clocks at optical frequencies only became possible with the advent of frequency measurement technology in the optical region, the mixing of laser frequencies in the broadband classical sense we were just talking about, a technology that originated from my MIT work here in the 1970s. An optical clock is a clock only if you can time its frequency.

There are already accurate optical clocks operating at a number of laboratories internationally. Optical clock technology has been used for a number of years to give unprecedented accuracy in high-resolution laser spectroscopy. Timing optical frequencies instead of the earlier wavelength measurements improved accuracies by tens of thousands of times. I am very impressed by the work that our French colleagues have done at the Universite Paris-Nord.

There have also been recent developments in laser-cooled trapped ion spectroscopy, capable of producing high-Q optical resonances for optical clock applications at accuracies higher than possible with Doppler-free resonances. Distinguished work has been done in our country at the National Institute for Standards and Technology [formerly the National Bureau of Standards], JILA [the Joint Institute for Laboratory Astrophysics, operated by NIST and the University of Colorado], and Bell Laboratories, as well as in Europe. Clocks with accuracy of one part in 10^{14}, 10^{15}, 10^{16}, 10^{17}, or even higher appear to be within reach.

One can see the blueprint of events to come in accurate measurement of time, perhaps in this decade or early in the next.

There are different approaches to measuring gravity waves, which could teach us many things about the cosmos. One possibility is to use extremely accurate clocks at optical frequencies in space. They would be spaced at distances approaching a sizeable fraction of the gravity-wave wavelength. Pulsars have periods of 10 to 100 milliseconds, so the wavelengths are very long. If you put highly precise and stable lasers into space, separated by about the distance from here to the Moon, for instance, they could detect gravity waves that would be hard to sense with terrestrial detectors. The terrestrial detectors do not have the luxury of relying on such long distances.

For some defense applications the possibility of having clocks in space is very attractive. Developing such a system would surely serve to advance the state of the art in optical clock technology, making possible fundamental experiments of the kind I have talked about here. My company is active in very precise optical clocks for the Navy and SDIO [Strategic Defense Initiative Organization], and has studied the uses of clocks in space. I hope that some of this work will be published soon.

With very precise optical clocks, it should be possible to detect in the laboratory the cosmological expansion of the universe. From the Hubble constant, all distances in the universe expand by about one part in 10^{10} in one year. This expansion can be detected on earth by observing the way two different clocks of the right designs keep time.

I like to think in terms of expanding the frontiers of knowledge with these fundamental experiments. However, it is no longer possible to do substantial work of that sort just by taking ordinary laboratory lasers. To expand the frontier of knowledge, one must invest heavily in advancing the technology. If defense needs can justify the expenditure of funds to advance the technology, fine, I'm all for it. Once the technology is advanced, it can be used to expand the frontiers of knowledge.

Time is the most fundamental element in nature. Time and space together form a four-dimensional continuum, an absolute reference frame for everything in nature, but I consider the time component of this continuum more fundamental in some ways, just because we can measure it more accurately. Any new approach to measuring time at higher precision has possibilities of new applications in technology, and in science. That's an important area, and I feel there's a great deal in store for all of us to watch for.

An earlier version was published in *Lasers and Optronics* ®
(formerly Lasers and Applications) a Gordon Publications, Inc. publication.

ROBERT N. HALL

The Semiconductor Laser

Robert N. Hall spent his entire career in semiconductor research at the General Electric Research & Development Center in Schenectady, N.Y., which he joined after completing a doctorate in nuclear physics at the California Institute of Technology in 1948. In 1962 he conceived of a way to obtain laser emission from a semiconductor junction and assembled a team at GE that demonstrated the first gallium-arsenide injection lasers within a few months. Almost simultaneously a group at the IBM T. J. Watson Research Center in Yorktown Heights, N.Y., reported stimulated emission from GaAs devices that were similar, but lacked the Fabry–Perot cavity used for mode selection at GE. In a matter of weeks, a third independent group at MIT's Lincoln Laboratory also reported GaAs diode lasers. Those early devices required cryogenic cooling even for pulsed operation. Continous-wave, room-temperature operation of a diode laser was not demonstrated until 1970, long after Hall had turned his attention to other semiconductor research.

Hall was awarded a Coolidge fellowship from GE in 1970. He retired in 1987 and remains a consultant to the company. He is a

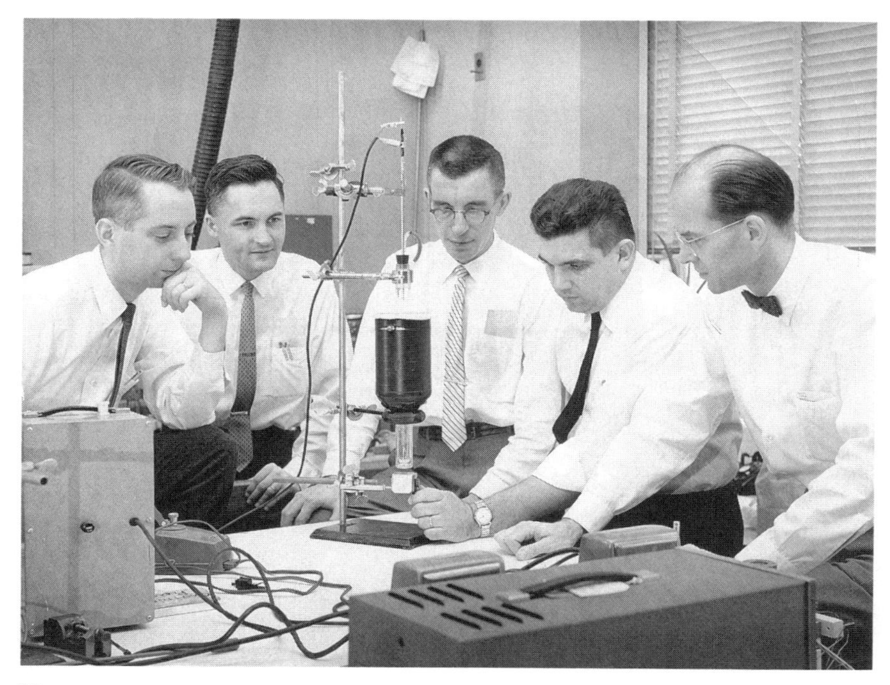

Figure 9.1 The General Electric team that made the first semiconductor laser. From left to right, Jack Kingsley, Dick Carlson, Gunther Fenner, Ted Soltys, and Robert Hall (courtesy of General Electric Research & Development Center).

fellow of the American Physical Society and the Institute of Electrical and Electronic Engineers, and received the David Sarnoff Award from IEEE in 1963. Jeff Hecht conducted this interview on October 11, 1984, in Hall's office at the General Electric Research & Development Center.

Q: How did you first get interested in science?

Hall: When I was young, we lived in Puerto Rico. We were somewhat isolated, and I needed things to do. We had a set of the *Book of Knowledge* which had a section on "Things to make and do." We did a lot of those things; I remember learning about making gunpowder and making a cannon and blowing various things up. The real start was when I came to live with my grandparents in New Haven for junior high school. My uncle, who was an inventor, took me to a science and technology fair in New Haven, and I saw a lot of things that just astonished me. I saw ball bearings bouncing off steel plates through hoops, and a tin-can motor that was just spinning all by itself, though there was a three-phase rotating field under the table. I *had* to learn how to do that. I got interested in motors, went to the library, and learned how to make a little motor with a commutator that actually worked. So my interest started with that science fair and maybe a natural curiosity. Certainly a big factor in getting me started was nudging from my uncle, who helped me to understand Ohm's law, and explained things, and showed me where books were in the library.

Q: Where did you go to school?

Hall: I went to high school in Alameda, Calif.—then I went to Caltech. I graduated in 1942, and came to General Electric.

Q: What were you originally doing here at GE?

Hall: I came here on a test program. I started in the research laboratory, and technical staff was very short—it was early in the war—so they asked me to stay on. It was a good place to work, so I stayed, working on continuouswave magnetrons for jammers and other microwave components. I also did a little work on special germanium diodes which Harper North was developing here. They were intended for microwave mixers, but he found that he could make very high voltage point-contact rectifiers.

Q: That must have been very early in semiconductor research.

Hall: It was before anybody ever made a transistor. People were just playing with germanium, silicon, and other semiconductors, mostly for detectors and mixers for radios.

Q: Then after the war you went back to graduate school?

Hall: Yes, I went back to Caltech and got a doctorate in nuclear physics. I came back to GE because I liked the environment and the people. Shortly afterwards, the transistor was announced by Bell Labs. I was talked into looking into it a little bit and found it very interesting. This was in the early days of semiconductors, and it was a very fruitful field.

Q: How did that lead you to the semiconductor laser?

Hall: That is quite a long story. There had been some semiconductor work here, and it seemed to me that the principal problem was in getting sufficiently pure germanium. So I devoted a year or two to making very high purity crystals by fractional crystallization, which turned out to be very efficient. We got into pulling single crystals of germanium following the work at Bell Labs. We found we could make rectifiers by alloying and regrowth, and this gave us a way of making a *p-i-n* rectifier, which has become very well established as a high-power device. We worked on an alloying process for making transistors, then came a period when we were on tunnel diodes because Leo Esaki had announced this negative-resistance phenomenon from extremely heavily doped junctions. We explored tunnel diodes in germanium and silicon, and in III-V compounds. This fortunately gave us a good handle on making very heavily doped III-V compounds, gallium arsenide in particular. So we had in-house people—myself and fellows working with me—who knew how to build things from gallium arsenide with very heavily doped junctions. These are the same process steps you need to make a semiconductor laser. So we were sort of ready to invent the semiconductor laser once we thought about it.

Q: At the time, weren't there a number of people talking about electroluminescence and III-V compounds?

Hall: Most of the III-Vs are direct-transition semiconductors, so radiative recombination is much more efficient than it is in germanium and silicon. The ruby laser and the gas laser had just been made, and lasers were an exciting field. People had written some articles speculating on ways to make semiconductor lasers.

Q: Who were some of those people?

Hall: [Nikolai] Basov in Russia had done a lot of early theoretical papers, and Benjamin Lax at MIT. Maurice Bernard in France had done some very interesting work. It explained in terms that I could understand what was meant by a population inversion in a semiconductor. I was a stranger to lasers and first had to learn some of the principles. I remember that people had occasionally asked me whether a semiconductor laser was possible. I originally pooh-poohed the idea, but those questions made me look into it and learn what the requirements might be. I remember looking through some conference proceedings and thinking that it didn't seem those ideas would work.

Q: What changed your mind?

Hall: The thing that really set it off was going to a device conference and hearing a paper by [R.J.] Keyes and [T.M.] Quist [of MIT Lincoln Laboratory]. I believe there was also a similar paper by Jacques Pankove from RCA. They talked about very high intensity radiation from a GaAs *p-n* junction, which was really astonishing. At high current densities, they claimed that something on the order of a kilowatt per square centimeter of light was coming from those junctions.

Q: Was that the first LED?

Hall: No, there were LEDs being made, but this was the first indication that they could be very efficient. Previously we thought maybe 0.01% efficiency might be reasonable, and here they are talking close to 100%. That really shook me up; that was the light bulb turning on. If things are really this efficient, what can you do with it? Now those thoughts about making semiconductor lasers began to make sense. With this evidence of very high efficiency and high power densities, I began putting numbers down. I remember on the train ride back from the meeting thinking about this and trying to write down numbers. It looked as though you could get a population inversion, and then the thing ought to work, if you could build the right kind of structure. So I began thinking how you might make a junction, include a cavity somehow, and get the Fabry-Perot geometry that would sort out the single mode you wanted and get coherent light coming out.

Q: That is what you brought home?

Hall: Yes. I did some more analytical work, putting numbers together to see how they turned out. They looked very encouraging, so I thought about possible structures and worked out the idea of making a *p-n* junction, cutting it up, and polishing. I had been an amateur astronomer and back in high school had made my own telescope, so I knew about ways of handling optics and polishing so you could make little structures suitable for a laser. Nowadays you cleave lasers, but we did not know about that then.

Q: How did the project get started?

Hall: We had a good deal of freedom to move from one project to another. I figured we would need half a dozen or so people and talked to a number of people who I thought might be able to contribute. They sounded interested, so I went to my boss, Roy Apker, and told him that I had some fellows rounded up and thought the idea might work, and asked how about having a go at it. He gave us the okay, so we went to work.

We had a very enthusiastic crew. Ted Soltys did the fabricating; he had been making GaAs tunnel diodes, and I suggested ways for him to make the junctions, so he got to work. Gunther Fenner was very good at electronics, so he began setting up equipment to pulse the diodes. We knew we had to get as high a current density as possible, and that to drive them that hard without burning them out, you had to give a quick pulse, then look for any light coming out.

We did not know what might come out, so we thought of various tests, and got an imaging tube to convert the infrared GaAs radiation into the visible so we could see the radiation pattern and watch for some kind of change when the emission went coherent. We thought we would see a change in the current-voltage characteristic, but that never materialized. Jack Kingsley had been working on lasers and was familiar with the use of spectrometers to look for line narrowing. Dick Carlson also helped us with some material preparation.

Figure 9.2 Hall in 1962 with one of the first semiconductor lasers made at General Electric, which was suspended in liquid nitrogen (courtesy of General Electric Research & Development Center).

We expected that our first experiments probably wouldn't work. We tried what we thought might work, but meanwhile we thought we would have to do some more serious basic work, so Carlson was working on materials problems associated with preparing heavily doped junctions. We made some junctions, got them assembled, and Fenner had his apparatus going to test them. He came in weekends sometimes, and one Sunday afternoon he gave Roy Apker a frantic call saying there was something going on that he did not understand, but that something was happening! We all got excited, and on Monday we all got together and looked at it. We were never able to reproduce the original thing he saw, a strange bar of light across a Polaroid of the image tube. But by Monday he had some things that looked more sensible, and we could see clearly that there were some very interesting patterns. Above a certain threshold current, the spectrum changed very drastically and you could see intensity patterns and modes showing up.

We knew we had something going for us then. It was a big rush from that point on. Some diodes worked, but of course most didn't.

Some did very strange things that we couldn't make much sense out of, but a few behaved in ways that we could understand and interpret as clear evidence for coherent light emission. Jack Kingsley got his spectrometer going and saw very clear spectral narrowing and multiple modes at the Fabry-Perot frequency spacing that we expected. We could see radiation patterns consistent with the number of lateral modes across the face of the laser, and taking into account the dimensions and wavelength of light, we could understand the far-field pattern. So we got enough to be convincing, and put it into a letter to *Physical Review Letters*. It came out on the same date the IBM group (M.I. Nathan, W.P. Dumke, C. Burns, F.H. Dill Jr., and G.J. Lasher) published their paper in *Applied Physics Letters*.

Q: And just after that Lincoln Lab group [Quist, R.H. Rediker, and Keyes] had a paper in *Applied Physics Letters*. How long had you been working on semiconductor lasers?

Hall: It was a very short time. The device conference was in June, and the publication date was November 1. [The journal received the paper on September 24].

Q: Had IBM and Lincoln Labs been working on the problem longer?

Hall: I believe so. They had been very interested in coherent emission and did some very worthwhile work on pointing out the importance of a direct-transition semiconductor, such as GaAs, for efficient conversion of electricity to light. I think they looked at the possibility of a laser but didn't describe how it could be done. I'm not sure if they were actually working toward a laser. [IBM did have a contract from the Army Signal Corps to work on a semiconductor laser—*Ed*] They were studying luminescence from GaAs diodes, pulsing them harder and harder. In a few cases they noticed spectral narrowing and knew they were getting stimulated emission, which is part-way toward coherent light from a laser. That was the result [the IBM group] published.

It wasn't the same structure we had; our laser had a Fabry-Perot cavity, which you could say was a proper structure aimed at making a semiconductor laser. They did not have anything that would give them mode selection, but they knew they had something very close to a laser. It wasn't long after that—I suspect after they saw our paper—that they recognized they needed a reflective cavity around it.

Another point is that most papers that had speculated about semiconductor lasers had indicated that the mirrors would be parallel to the junction plane, so the radiation would go perpendicular to the junction. The thing that made ours work, and the basis for all subsequent semiconductor lasers [until the recent development of vertical cavity semiconductor lasers—*Ed.*], was arranging the Fabry-Perot mirrors so the radiation would bounce back and forth in the junction plane. This gave a relatively long path for amplification. Also, right from the start I recognized that if the radiation was going perpendicular to the junction you would have an awful lot of losses from absorption by the heavily doped material on both sides of the junction.

Q: Had you been aware of others working on semiconductor lasers?

Hall: We had to guess. We knew that Keyes, Quist, and Rediker were because they showed their work on high-efficiency luminescence. And we knew that Pankove probably was because he had given a similar paper. We knew that IBM had thought about the possibility but had seen nothing to indicate they were actively pursuing it. Their letter took us totally by surprise.

Q: Did the others know about you?

Hall: No, I think we caught them totally by surprise. We had done tunnel-diode work, which in a way is a close step because it involves degenerately doped semiconductors and gallium arsenide. So we had all the ingredients but had never done anything on optical emission that I can recall.

Q: Didn't you have some interaction with Bernard?

Hall: He came to our laboratory several times, discussing the possibility of semiconductor lasers. He appeared when we had one going but before we had submitted our paper. I felt a little bit awkward trying to discuss the problems of how you might go about it and what the problems were, and not being able to tell him that we had one in the next room.

Q: Why do you think you were first?

Hall: The device conference provided an immediate stimulus to all of us. I guess we all had an equal chance then, and MIT and RCA probably had a head start. But we happened to have people with the right talents who could be put together and who weren't tied to other projects, and a manager who said go to it. We had a loose and informal organization, with people who knew how to work together and could do it very efficiently and quickly.

Q: Then the environment at GE helped make it possible?

Hall: Yes. Also, we happened to recognize the necessary ingredients; some thoughtful design went into it to begin with.

Q: Were you the manager or leader of this group?

Hall: I have never been a manager, but I guess I could be called the leader. It was mainly my idea, and the other fellows all pitched in and worked hard.

Q: What were your hardest technical problems?

Hall: First, we had to get some good gallium arsenide, which was hard to find then, though we had an assortment of material on hand from the tunnel diodes. We had to cook up ways of diffusing junctions. A lot of the GaAs was just dead, but in some cases we found junctions that would light up pretty well.

Another problem was making the structures themselves. I learned how to cut, lap, and polish these things, and to check the parallelism of the two sides, which was tricky to achieve. Ted Soltys mounted them on headers with good husky leads so they could handle high currents. Gunther Fenner set up the electronics to get very high current pulses and set up the liquid air flasks so you could look at the light coming out. It wasn't anything you would call a scientific challenge, but it was an engineering and technical challenge to assemble this equipment in a short time and get it running.

Figure 9.3 Hall today (courtesy of General Electric Research & Development Center).

Q: What more did you do with semiconductor lasers?

Hall: We followed the field for a while, trying to understand the mode patterns and trying to get more efficient lasers. Then we began to think about possible applications, which you could see once you learned something about their properties. They seemed to be mainly in communications, where GE had relatively little interest, so we didn't see major applications for semiconductor lasers within the company. Lots of other things did need attention, so after about a year or so most of us trickled off, and the semiconductor laser work just stopped.

Q: Was that a management decision?

Hall: I wasn't under pressure to drop it. If there was a clear need in the company, operating components would be willing to pick it up, but we didn't have that kind of encouragement.

Q: The need for cryogenic cooling obviously was a major drawback of early semiconductor lasers. Why did it take so long to get away from that?

Hall: You had to make very efficient heat sinks and very small structures so the heat could spread out laterally. In our early lasers, the junction was across the entire plane of the device. To go to higher temperatures, you need stripe lasers, so the current density will be very high just in a very narrow region where it is needed, and the dissipated power is as little as possible so the heat could spread out laterally and into the heat sink. They had to develop copper and diamond heat sinks and just engineer the dickens out of this problem of getting the heat out. It took quite a few years, with a lot of tough engineering steps.

Q: What are you doing now?

Hall: I have been looking at problems relating to very large scale integrated circuits in silicon (VLSI), which GE is putting a big emphasis on. One of my long-term interests has been in semiconductor defects and possible impurity contamination, so I have been looking into some of our processing steps to see to what extent impurities are problems.

Q: Are you surprised at how far semiconductor laser technology has come?

Hall: Yes, I am very much surprised, or maybe not so much surprised as impressed. Although I'm not actively involved with semiconductor lasers, I am familiar enough with the field to recognize that an awful lot of tough engineering problems have had to be worked out. It has been very interesting watching this develop through all the painful steps to the point where they have very efficient singlemode lasers working at room temperature, tailored to the right wavelength. It is just a marvel the way it has grown, and the way the various formidable problems have been chopped off, one at a time.

Q: Where do you think the technology might go?

Hall: I suppose there will be lots of little gadgets, in cameras or home recorders or communications. I expect fiber-optic communications will crop up all over. I suspect that the semiconductor laser is so small it will crop up in applications you really can't predict. But I guess I have let that world go by. I do not so much follow the instrumentation end of things as I do try to understand the basic physics of semiconductors. There are plenty of mysteries down at the microscopic end of things to keep me occupied.

Q: If you had it all to do over again, would you still work on the semiconductor laser?

Hall: I don't see any other way of answering that than yes. That was an experience I really treasure, a classical model for inventing something. You get some ideas, you put them together, the light bulb turns on and suddenly you see a way of doing it. You go at it as hard as you can, you get some guys working with you that you like to work with. The unexpected thing of this early success made it all the better. There are a few such things that happen in a career, but not many.

Q: It is hard to find encores for that kind of thing. Do you regret having left semiconductor lasers?

Hall: No, in fact I feel a little bit, not quite ashamed of myself, but as though I had abandoned the field, and had all the fun of discovery without sweating out the hard work of, you might say, rearing the baby.

An earlier version was published in *Lasers and Optronics* ®
(formerly Lasers and Applications) a Gordon Publications, Inc. publication.

C. KUMAR N. PATEL

The Carbon-Dioxide Laser

A native of India, C. Kumar N. Patel came to the United States for graduate study in electrical engineering at Stanford University. After receiving his doctorate in 1961, he joined Bell Laboratories. He has become increasingly involved in management, and since 1981 has been executive director for research of the physics division at AT&T Bell Labs, Murray Hill, N.J.

Patel's pioneering work on molecular gas lasers included discovery of the 10-micrometer carbon-dioxide laser and the 5-μm carbon-monoxide laser, and demonstration of high-power output from CO_2. He also developed the spin-flip Raman laser, which he used to measure nitric oxide concentration in the stratosphere in pioneering experiments that played an important role in the debate over the fate of the supersonic transport aircraft in the United States. Other research interests have included absorption measurements in highly transparent materials, spectroscopy of solid hydrogen, and medical uses of CO_2 lasers.

A member of the National Academy of Sciences and the National Academy of Engineering, Patel is also a foreign fellow of the Indian

Figure 10.1 C. Kumar N. Patel watches an early continuous-wave carbon-dioxide laser at Bell Labs in 1965 which generated a then-record output of 106 watts (courtesy of AT&T Bell Laboratories).

National Science Academy. He has received numerous awards, including the 1966 Adolph Lamb Medal from the Optical Society of America, the 1976 Zworkin Award from the National Academy of Engineering, the Institute of Electrical and Electronics Engineers' Lamme Medal in 1976, Texas Instrumentation Foundation's Founders' prize in 1978, and OSA's Charles Hard Townes Medal in 1982.

Jeff Hecht conducted this interview on November 6, 1984, in Murray Hill, N.J. It was updated in July 1990.

Q: How did you get involved with physics and lasers?

Patel: I did not start out to be a physicist. I had planned to go into the Indian Foreign Service after I got my bachelor's degree in engineering. However, you had to be 22 to take the entrance examination, and I was three years too young when I graduated. I wanted to do something useful during those three years, so I decided to get a PhD. After the first year at Stanford, I realized I was having so much fun doing research in physics that I wasn't going to go back and take the foreign service exam.

Q: What was your doctoral research?

Patel: I studied ferrimagnetic resonance in yttrium iron garnet and worked on a narrowband microwave filter under Dean Watkins, who soon afterwards left Stanford to devote full time to the Watkins-Johnson Co. I finished up in early 1961 and came to Bell Labs. My supervisor asked what I wanted to do, the standard question you ask a new PhD at Bell Labs. I had had enough of microwaves, so I said I wanted to do high-resolution spectroscopy with lasers. I didn't know much about lasers or spectroscopy, but it sounded interesting. This fellow was quite savvy, and he suggested I start with lasers and see where that would go.

Q: How aware of lasers were you then?

Patel: I had seen newspaper accounts of Ted Maiman's ruby laser, and read Ali Javan's paper on the helium-neon laser in *Physical Review Letters.* Arthur Schawlow had had an article in *Scientific American.* But that was about all. One virtue of this institution is that it lets people get into fields where they have no prior knowledge. All they require is enough basic understanding of science.

Q: Then they basically turned you loose?

Patel: That's right. My supervisor pointed me to people like Ali Javan and Bill Bennett, to bounce ideas off. The large collection of people and accomplishments is one strength of this place. But another is that it's so big that it's impossible for a newcomer to talk to too many experts, because there are so many of them, and you don't know who they are.

The basic naivete of an individual who comes into a new field is a tremendous advantage in getting things done, because he does not know what some old fogeys have decided is impossible. I know how I decide now what is not possible today. The trouble is that two years from now I just remember the decision, not the arguments I used to arrive at that decision. Things will have changed in two years, but I will remember my conclusion, not the caveats that went with it.

Q: Did you have any particular ideas in laser spectroscopy?

Patel: I really couldn't have done any spectroscopy because there were no tunable lasers. There were just a few lasers then, mainly ruby and helium-neon, so the first thing to do was to try to make new types. We were being taught that helium-neon was an outstanding system because helium transferred its energy to neon through selective excitation and a population inversion. I wondered if you could get away from those two-gas systems. If you could get the inversion by an incoherent excitation mechanism such as a discharge, without needing an intermediate gas to transfer the excitation, then it seemed there would be an enormous number of possible laser systems.

The major breakthrough came within six months after I got here. I showed that you did not need helium in a helium-neon laser. It did not work quite as well as the helium-neon system, but it proved you did not have to have that kind of energy transfer. Of course, a few years later somebody showed that you could put almost anything in a discharge tube and get a laser if you hit it hard enough. That was disconcerting at the time.

Q: Ron Waynant once called that the "telephone pole" theory of lasing, the idea that if you hit a telephone pole hard enough it would lase.

Patel: Schawlow and one of his students demonstrated that dramatically in the mid-sixties when they fired a ruby laser into a bowl of Jell-O. There was a bit of cheating in that it was doped with Rhodamine 6G, not the kind of Jell-O you would want to eat, but it was close enough. However, before we could get to that, we had to learn how to walk.

I had started looking for a tunable laser for spectroscopy. The next best thing was to have laser lines close enough to each other, so you could do point-by-point spectroscopy. I thought a large number of lines would let us do that.

I did have another approach to spectroscopy. One of the earliest high-resolution laser spectroscopy experiments was one I did using a xenon laser. I shifted the 2.02-micrometer wavelength by applying a magnetic field to the discharge, and was able to map out the Doppler profile of xenon in another tube. It wasn't tunable by a lot, but it was tunable enough to use. Before other tunable sources came along, some of my colleagues used magnetically tunable lasers for limited applications in spectroscopy.

This was about 1963, and I was getting much less excited about just looking for another transition. Gas lasers typically produced milliwatts of power. Someone here had built a 15-meter long helium-neon laser, which put out about 150 milliwatts, but that size was too much. A lot of people were saying that gas lasers were good as little laser light bulbs for drawing straight lines, but that if you wanted high power you needed either ruby, neodymium-glass, or neodymium-YAG.

I changed my tack and asked if there might be something basically wrong in looking at rare gases. Of course, at room temperature only rare gases are atomic gases; everything else is molecular. The problem with atomic systems was that the first excited state is far above the ground state, so you have big losses.

I started looking at the possibility of using molecular systems. Carbon-dioxide was the first system I looked at seriously, and it had lots of advantages. You have to get away from electronic states, because they give you the same problem as in atomic gases—the levels are too far above the ground state. Diatomic molecules such as carbon monoxide have just a single ladder of vibrational states, and you can show from simple quantum mechanics that the vibrational level lifetime gets shorter for higher vibrational states. That isn't what you want in a laser candidate, so you have to go to at least a three-atom molecule. There you could have different vibrational ladders, and if you are lucky you can get a system where one vibrational mode has a longer lifetime than a lower-energy one.

Having done those calculations, it was quite clear that CO_2 could work. It did the first time we tried. You knew where to look for the radiation, near 10 micrometers, and by golly, it worked marvelously well. We got tens of milliwatts on the first shot.

Figure 10.2 Patel with a flowing-gas laser at Bell Labs in 1967 (courtesy of AT&T Bell Laboratories).

Q: Were you surprised that the CO_2 laser worked so readily?

Patel: Yes. I did not think I knew enough about CO_2 to have predicted it right the first time. Also, we were aware of Harry Boots' experiments in England, where most of the emission from molecular gases in a discharge was on atomic lines, so it wasn't entirely clear that you could keep the CO_2 molecules intact.

Q: How big was your device?

Patel: A meter and a half long. This work made us realize that you did not need a complete inversion between two vibrational levels to get lasing, because each vibrational level has a whole set of rotational states. Even when you have fewer molecules in the upper vibrational state than in the lower one, you still can have inversion between two rotational states.

The moment you realize that, everything becomes very clear. Then you say, why did I throw away the diatomic molecules? We went back and tried CO, and that worked too. I wouldn't have tried it first, because it didn't fit my preconceived notions of needing total inversion, but of course total inversion is not necessary.

Then I remembered about energy transfer in helium-neon, and wondered if there could be something like that in a molecular system. That led to the nitrogen carbon-dioxide laser system. Molecular nitrogen is very strange; if you throw energy into it, the energy just stays there, with a lifetime of seconds. Nitrogen's first excited vibrational energy level is the same as that of the CO_2 upper state. You put the two things together and the next day you get 10 watts from the same tube that gave you 10 mW before. I think that was the highest continuous-wave power that anybody had had. Within six months we had destroyed all the earlier myths about where you could get large amounts of laser power.

By about mid-1964 it became clear that de-excitation wasn't really complete in carbon dioxide. Then the game was to find something else to add to the system to drop that lower level of CO_2. We tried water vapor, an idea that came from work on rocket engines. We also realized helium was good, because it has enormous heat conductivity at pressures of 10 or 20 torr. It turned out that the best mixture is carbon dioxide, nitrogen, and helium. One can add a few other things here or there, depending on whether or not you seal the tube, but beyond that I don't think there have been major advances in the gas mixture. By mid-1965 I had a 200-watt continuous-wave CO_2 laser, which was more than enough power for anything you wanted to do in the laboratory.

Q: Why did you move on from CO_2?

Patel: Gas laser work wasn't providing the same kind of excitement. I had had the field to myself until we published the paper on the 200-W laser, but then it began getting crowded. I try to stay away from where most of the people are, because I think the fun is most likely where nobody else has tried anything. Nonlinear optics was just becoming fashionable. Joe Giordmaine and R.C. Miller had just operated the first parametric oscillator. That gave me the germ of an idea for a tunable laser source using a carbon-dioxide laser. All we needed was nonlinear optics farther in the infrared than anybody had done before.

The whole question of nonlinear optics for the infrared was open. The first material I tried was tellurium, which turns out to have the largest known nonlinear coefficient in a phase matchable material. We stuck the sample in, and it generated the second harmonic of CO_2 exactly the way we predicted it would. The problem of making a parametric oscillator in tellurium turned out to be a lot more difficult than I had thought. Multiphoton excitation of electron-hole pairs in the semiconductor generates enough added absorption to overcome the gain.

About this time, 1966, Dick Slusher came to Bell Labs. He wanted to use the carbon-dioxide laser to study Raman scattering in semiconductors. His supervisor, Peter Wolff, had done a theoretical study that showed that Raman scattering from electrons in a semiconductor placed in a magnetic field could give you light output tunable in frequency. We got spontaneous light scattering the first time we tried it, using indium antimonide. The spontaneous Raman scattering was a very broad line, incoherent and very weak, coming out in all directions.

The money was in making a Raman laser, but the problem turned out to be very hard. Slusher got interested in other things, but I kept working on it. By this time Earl Shaw had joined my effort and eventually the spin-flip Raman laser lased in 1969. There had been Raman lasers before, but this was the first tunable Raman laser at any frequency.

Q: What made it so hard?

Patel: We knew very little from theory or experiments about the nature of the Raman scattering process. We had fallen into a frame of mind that said we didn't have enough Raman scattering, and Raman scattering is proportional to the number of electrons, so let's increase the number of electrons. Things got worse, not better. It took me a year and a half to realize that if you have too many electrons, both the upper and lower levels of the Raman transition will get filled, and once they're both completely filled, you can't get stimulated Rarnan scattering. When we recognized that, the spin-flip Raman laser worked within two weeks.

Q: So at last you had a tunable laser?

Patel: Yes, I had reached a point where spectroscopy was possible. For a number of reasons, the indium-antimonide spin-flip Raman laser hasn't turned out to be practical, but we used it for five or six years as a workhorse for a variety of things. Soon after my work, Lincoln Laboratory pumped indium antimonide with a carbon-monoxide laser, which enhanced the cross-section because the pump frequency was close to the 5.3-μm bandgap. They also showed that the spin-flip Raman laser could emit continuous-wave; our initial work was pulsed. I too jumped in and showed that going to InSb samples with high electron concentration actually made the spin-flip Raman laser threshold go down because the spontaneous Raman scattering linewidth gets smaller. Also now you could use simple electromagnets or even permanent magnets for tuning the spin-flip Raman laser rather than the superconducting magnets used earlier.

We used spin-flip Raman laser spectroscopy for pollution detection, to see very small concentrations of gases. Lloyd Kreuger and I did the earliest tunable laser optoacoustic spectroscopy experiments. Optoacoustic spectroscopy was very useful because instead of trying to measure transmission through a sample, you try to measure the energy left behind. We showed you could measure absorption coefficients as small as 10^{-10} inverse centimeters in a sample length of 10 centimeters.

I spent two or three years looking at a variety of pollutants including nitric oxide. Nitric oxide was rather interesting because there was tremendous sensitivity about what was coming out of automobile tailpipes, and nitric oxide was one of the constituents which was rather nasty. We showed that one measurement apparatus could measure concentrations from something around one part per billion to one part per thousand, implying a tremendous dynamic range.

The real excitement came when I decided that the right place for this kind of measurement was not on the ground. People were worrying about ozone depletion in the stratosphere, primarily from supersonic transports. Johnson at Berkeley had a model which said that if there were 400 SST flights every day, ozone concentration would drop significantly in two years, and this would increase the ultraviolet coming down. But nobody had measured the concentration of the nitric oxide that was supposed to catalyze the reaction.

We wanted to measure nitric oxide concentration at 28 kilometers. You can't take samples up there and bring them down, because by the time they get down, it will all be nitrogen dioxide. You want to put an apparatus up there before sunrise, and let it go through the whole day so you have an idea of the variation through a whole diurnal cycle. We and Sandia Laboratories built an apparatus and lifted it up with a balloon to 28 km. We found that the model had to be corrected by a small amount, and Johnson's calculations were basically verified by actual measurements. However, I am not sure if the nitric oxide problem in the end was part of the justification for scrapping the US SST program.

Having gotten to this stage, I had fulfilled in some sense what I wanted to do when I came to Bell Labs—high-resolution spectroscopy with lasers. It took me 14 years, but in the meantime lots of other things fell out of it. One of them was studies of hydrogen cyanide production from the platinum catalysts used to clean up automobile exhaust. Some of our work led the Environmental Protection Agency to rewrite catalyst specifications to mandate that the HCN coming from the tailpipe could not exceed some acceptable level.

Q: Have you looked at other materials?

Patel: I have been looking at very transparent solids, mostly in the visible and the near infrared. We find there is no such thing as a totally transparent material; things absorb somewhere, and these small absorptions turn out to be important. We've been looking at the absorption coefficient of water, which people have done for at least a hundred years, and there is roughly a factor of 20 variation in reported data.

Q: That large a variation for pure water?

Patel: As pure as can be. Whenever I find something like this, I consider the field ripe for doing something worthwhile. When you look at the problem carefully, you realize that the variations arise because the absorption coefficient is small, and you have to use very long paths if you use conventional measurement techniques. The optoacoustic technique lets us avoid seeing window absorption, and make measurements with absolute accuracy ±10%. We also found the absorption minimum is at about 4700 angstroms, not at 4900 or 5000 angstroms where most people used to think it was. That is important for satellite-to-submarine laser communications.

Q: Spectroscopy also is used in more research-oriented fields. Have you been working in those areas as well?

Patel: That's what I'm doing today, looking at solid molecular hydrogen. We don't think there is any immediate practical impact, at least today, but there is some really interesting physics. Most solids are atomic, but hydrogen is a molecular solid. The hydrogen molecule sitting in a solid lattice still behaves as if it were a free hydrogen molecule. Because hydrogen atoms are so light, their vibrations cover much of the distance between two molecules, and the tremendous amount of overlap of wave functions between the molecules brings into focus lots of really exciting physics.

Q: Aren't you also heavily involved in management?

Patel: Right now I don't get paid for doing research, I get paid for management. The research I do is what I call "for the soul." My boss never asks me how many papers I publish or what research I am doing. My getting into management is very similar to my having fallen into physics. I generally enjoy it and get a kick out of it. I don't think I would enjoy it if it were strictly what one might call managing people, but in the research area it is technical management, management of ideas.

Q: How do you see your role?

Patel: In predicting the next breakthrough and putting our resources there. There is much more technical involvement than in managing other kinds of enterprises. I do very little of what is traditionally called management. I do no financial management— I have a specialist responsible for that. I do no personnel management; the specialist does that. My task is to make sure that the money that is entrusted to me by AT&T will create the needed breakthroughs, which means I have to find out where I am going to put my money.

Q: The investment in the sort of laser research you did is now paying off in technology. Where do you think the CO_2 laser is going?

Patel: CO_2 lasers are becoming more compact, more powerful, more efficient, and more user-friendly. Engineers are considering the CO_2 laser as part of a system rather than a free-standing entity. The application base and the customer base for CO_2 lasers continues to expand to include new applications and new customers. Most industrial applications of CO_2 lasers have involved material removal or welding. New applications that build up materials are now becoming possible, completing the spectrum of materials processing. Such materials-buildup systems are facilitating rapid prototyping of complex plastic and ceramic shapes directly from computer generated data.

Offshore manufacturers are beginning to make inroads into the U.S. and international markets for CO_2 lasers. I believe that competition based on price and performance will become even more common. The features that CO_2 laser suppliers can offer to differentiate their products are going to be user-friendliness, computer controllability, and flexibility.

I am somewhat disappointed that progress in fibers for 10.6 μm has essentially become stalled. Yet the bright side is that applications of CO_2 lasers in all fields have continued to expand. With the increasing ubiquity of CO_2 lasers in materials processing, we will see "garage" operations that provide such services the same way that "garage" operations provide traditional machining services. Another very pleasing aspect of the CO_2 laser field is that in spite of the enormous technological base for the laser and its applications, scientific publications continue unabated on CO_2 and other molecular lasers and their applications.

Once again, we have proven that the advances in applications and technology of CO_2 lasers are driven by the value that they provide in the commercial arena rather than in defense-related activities.

Figure 10.3 Patel in his office in 1990 (courtesy of C. K. N. Patel).

Q: What about laser technology in general?

Patel: There are many uses in what I call information technology. In communications, the laser is simply a light bulb, and the question is how fast you can turn it off and on. Beyond that, there is optical logic, and switching in a totally optical environment. Some people say optical switching is here and now, and there are devices that are called optical transistors, but I don't think we are there yet, no matter what one calls it.

Q: What do you mean by that?

Patel: A transistor is a device in which a coherent or incoherent source of signal controls an incoherent source of power—what I call low-quality power. A switch that switches laser light with other laser light is not really a transistor, because the coherence and color of the light are more important features of what happens.

That's not to take away from the work on bistable optical gates, but the point is that for a practical computer you immediately run into problems with power consumption. I would like to see some switching technology that lets you go optical to optical, so you only have to convert signals to electronic when a customer wants to put it on his terminal or pick up his phone. It has a long way to go, more than five years, but I think it will happen. An optical computer is even further out because we are up against materials problems that have no simple solutions.

Q: If you had it to do over again, would you still work with lasers?

Patel: If I were starting today and forced to be a physicist, probably I would. But if I had a choice I probably would not do physics. There are fields such as biophysics and what people call "artificial intelligence" which are where physics was 20, 30, or 50 years ago. The fundamental understanding in those fields is very limited, so the possibility of making a significant advance for a unit of work is significantly higher than in lasers or physics.

Q: Do you have any regrets about working with lasers?

Patel: Absolutely not. I may be stealing somebody else's words, but it has been a wonderful ride.

An earlier version was published in *Lasers and Optronics* ®
(formerly Lasers and Applications) a Gordon Publications, Inc. publication.

WILLIAM BRIDGES

The Ion Laser

William B. Bridges was born in Inglewood, California, in 1934 and received bachelor's, master's, and doctoral degrees in electrical engineering from the University of California, Berkeley. He joined the technical staff at the Malibu Research Laboratories of the Hughes Aircraft Company in 1961.

After his first work at Hughes on microwave-tube research, he was caught up in the excitement of laser technology. Bridges discovered the argon ion laser in 1964, after two years of research on helium-neon and xenon lasers. For the next several years, most of his efforts centered on developing practical ion lasers for airborne applications. Throughout the early 1970s, his efforts concentrated on various aspects of carbon-dioxide lasers and their suitability for military systems applications.

Since 1977 he has been professor of electrical engineering and applied physics at the California Institute of Technology, Pasadena, California. In 1983 he was appointed Carl F Braun professor of engineering. Bridges is a member of the National Academy of Sciences and the National Academy of Engineering, and is a fellow

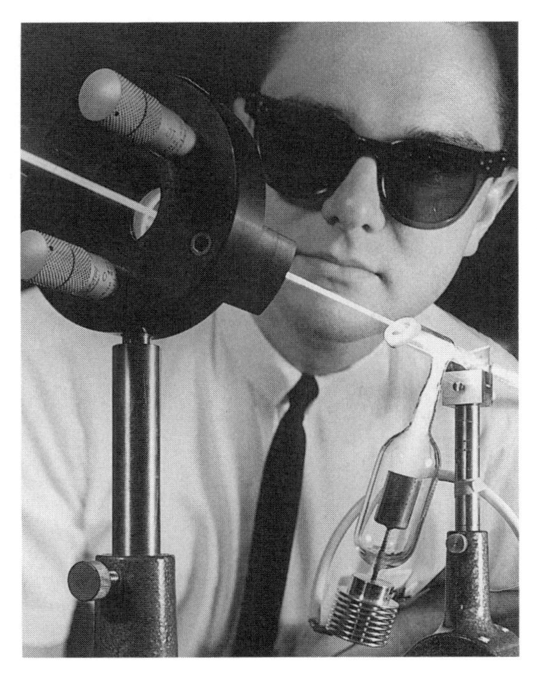

Figure 11.1 William B. Bridges poses with a pulsed argon ion laser in March 1964 at Hughes Research Labs, about two weeks after the laser was discovered. The picture shows the beam both inside and outside the laser cavity (courtesy of Hughes Research Laboratories).

of the Institute of Electrical and Electronics Engineers, the Optical Society of America, and the Laser Institute of America.

Richard Cunningham conducted this interview on November 9, 1984, at Caltech. It was updated in July 1990.

Q: How did you get involved in science in the first place, dating back to high school days or earlier?

Bridges: It seems like I have always been involved in science, although none of my family was involved. My mother worked for the local high school as a payroll clerk, and my dad was an auto mechanic. I suppose the only influence in that regard was my grandfather, who gave me the full run of his workshop, and a great uncle who did the same. I was always interested in building things.

Actually, as you know, I am an electrical engineer. I really got involved in radio in the grammar school era by building crystal sets. By the time I was a freshman in high school I was an active radio amateur. I built my own transmitter and antennas and all the equipment needed. Radio electronics was a common route of entry for people in my era, when "electronics" connoted something you built rather than something you bought from Japan.

Q: What were you doing professionally immediately prior to the ion laser work at Hughes?

Bridges: I had stayed on at Berkeley to do my graduate research in microwave electronics. My thesis work was on noise in electron beams of the kind used in traveling-wave tubes. Strangely enough, an effect that I discovered in my thesis work has recently resurfaced as a generator of high microwave powers: the so-called VIRCATOR, or *vir*tual *cat*hode *o*scillat*or*.

I joined Hughes Research Laboratories in December, 1960, although I think I only worked two weeks and then took a leave to go back and finish my thesis. I returned to Hughes full-time in June, 1961, and joined the Electron Dynamics Department. My section was involved in the research and development of microwave tubes, particularly in low-noise microwave and millimeter-wave tubes. This was a little more than a year after Ted Maiman demonstrated the first laser, so lasers were a hot item at Hughes.

My group was not involved in lasers specifically, at least not initially. But by about mid-1962 we had closed out an old low-noise tube contract. After seven years we couldn't think of any more to do, so we quit. Something you might not do these days is turn down money.

My supervisor then, Don Forster, was looking for some way to get into the laser excitement. In the early part of 1962 we thought of extending our microwave tube activity into very high-speed or broadband photodetectors. We started off by looking to see whether we could use use traveling-wave structures as photodetectors. This really followed some work that had been pioneered at Stanford by Tony Siegman and Burt McMurtry, his graduate student.

We actually built a couple of these things, and I worked cooperatively with Vik Evtuhov and Jim Neeland with their ruby laser setup at Hughes Research Labs. We used these high-speed detectors to puzzle out some of the funny mode properties of ruby lasers. Things were pretty crude in those days, and we weren't sure that ruby lasers were behaving as simple resonator theory would predict. It turns out that it's okay and theory wins. You just need a better quality of ruby to get the expected mode properties. But my first work with lasers was really in conjunction with the microwave photodetector end of things.

Earlier that year (1961), the helium-neon gas laser was first announced at Bell Laboratories, with the infrared line at 1.15 micrometers. At that point another group at Hughes started trying to build some helium-neon lasers following the Bell Labs design. We ordered one of their lasers for our phototube work. Well, the other group had some troubles, and one thing led to another. I got involved with this group, trying to get gas lasers from them.

About that time, the Bell people announced their red helium-neon laser in June, 1962. Mal Currie, who was then our associate lab director and now is president and chief executive officer at Hughes, said, "That's fantastic. Let's get into that." Overnight, I found myself in the gas laser business. The microwave phototubes were forgotten, and we were off and running with helium-neon lasers.

Q: What influenced you to start working on the noble gas ion area?

Bridges: From mid-1962 we worked on helium-neon lasers. Remember, Hughes was not really a commercial manufacturer in those days. We were building tubes for our own internal use and were trying to find appropriate military applications. We got into some of the other noble gas neutral lasers, particularly xenon. In fact, my first laser publication from Hughes was on an infrared xenon laser. It had high gain and the potential for operating in a straight-through amplifier mode. I spent about a year with that laser; a couple of publications evolved from it. In late 1963, when I was still working on this infrared laser, a very interesting publication appeared from Earl Bell and Arnold Bloom at Spectra-Physics on the mercury ion laser. It was interesting because it put out more power than some of the lasers we had seen before, and it was visible. They reported red and green lines in a short publication, which just described the wavelengths and some of the conditions of operation.

I had the freedom in those days to jump on things that were interesting, and that was the kind of thing you just can't leave alone. So Bob Hodge, my technician, and I put together one of these mercury ion lasers to see what it was like and what the pumping mechanisms might be. Bell and Bloom's original publication reported operation with a mixture of helium and mercury. I had done some work before with the helium-xenon laser and found that the commonly accepted idea of what role helium played turned out not to be true. So I wanted to see if a similar problem existed here. Maybe helium wasn't really necessary.

So we built one of these helium-mercury lasers, got it operating in the lab, and then started playing around with it to see what its characteristics were. Is helium essential or not? Is it really charge exchange? Are these Penning collisions? What's the pumping mechanism?

One way to answer these questions is to take the helium out and substitute neon. I had a nearly complete station of gases at hand so we used neon as a buffer gas—that works. Actually, the laser will work very well with pure mercury. Anyway, after we had made a neon-mercury laser, we decided we would try argon as a buffer gas to see if we could make an argon-mercury laser. Later, we found out that works, too. But this particular day we couldn't make it work. We were using much too much argon. Nor did we have it well adjusted.

So, we pumped out the tube, flushed it, and put helium back in to make sure that our mirrors were still aligned. To our surprise, we found we had a new line going in what was ostensibly a helium-mercury laser. We now had a blue line at 4880 angstroms in addition to the red and green lines from mercury. Well, that was very exciting and totally unexpected. It was a case of discovery rather than intentional construction . . . a nice Valentine's Day (1964) present!

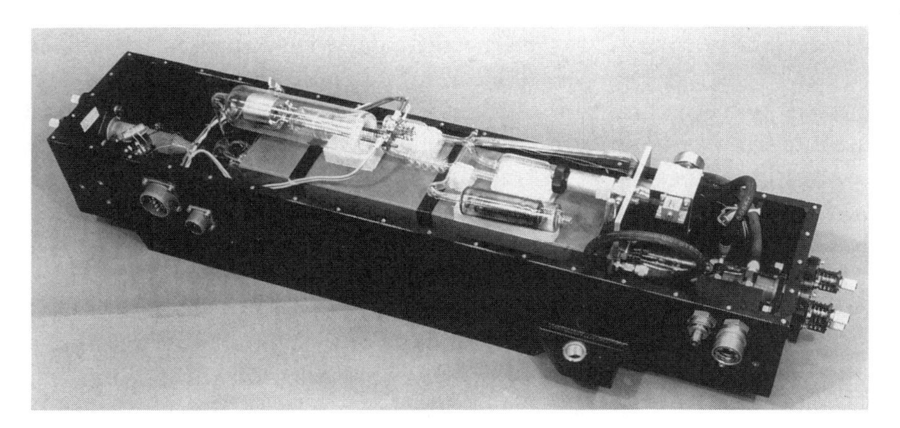

Figure 11.2 An airborne argon laser built at Hughes in 1965. The tube bore and envelope were fused silica; higher-temperature materials are used today (courtesy of Hughes Research Laboratories).

Q: That sounds like a combination of excitement and terror.

Bridges: I don't know if I can really convey the excitement. You see something unexpected, and furthermore you don't quite know how you produced it. Maybe it will go away, and you won't get it back again. To complicate things further, we didn't have spectroscopic-grade argon on the station. To do this experiment we wheeled up an old welding bottle with welding-quality argon and God only knows what possible contaminants. So we couldn't merely assume we were looking at an argon laser line.

We measured the wavelength to within a couple of angstroms with a small spectrometer we had. Then it was hurriedly off to the library while the laser was still running. I fumbled around in the library, because I was unfamiliar with the spectroscopy of ionized argon and didn't know where to look. But after about an hour, I concluded that we were looking at ionized argon.

Well, with that confidence, we felt we could start tampering with the laser. So we pumped it out, flushed it with helium again, and still had the line, but weaker. Several flushes finally made the line go away. We were now back to having a helium-mercury laser. Then we let a little argon back in and reproduced the blue line.

We had already made time-resolved measurements on the optical output pulse of the mercury laser. As I recall, Bloom and Bell didn't mention that the mercury lines lased only in the afterglow and didn't go with the current pulse. But in our laser, the argon line was very clearly going directly with the current pulse and the mercury lines turned on several microseconds after the current pulse cut off. This wasn't especially encouraging, because the currents and voltages we were using in this particular pulsed laser were hideously high compared with continuous-wave gas lasers of that era. But it seemed as though an argon laser might operate in a continuous-wave mode, although only at incredible inefficiencies.

Q: What happened to that first tube?

Bridges: The Smithsonian asked for the first ion laser not too long ago, but it had gone out in the trash the day after we discovered the argon line. We tried to get the mercury out of the tube so that we had just argon going. But once you get mercury in as a contaminant—and notice how quickly our thinking changed, considering that this had been a mercury laser just hours before—you can't freeze it out. We could almost make the mercury lines go away, but not completely. We decided that the only way to do this right was to start with a clean tube. So this one was cut off the station, the glass blower was given a rush order, and by the next week we had a clean tube running with just argon. That was even more exciting because now, instead of just the 4880 line, we quickly had 10 lines going in argon. That created a lot of activity around the laboratory!

Q: Was this the point at which you published your findings?

Bridges: Now we were stuck—should we publish immediately or make careful measurements and work things out? We really wanted to nail down the wavelengths very carefully to identify these lines. All we had in my lab was a very small monochromator; you could measure wavelengths to maybe within 1 angstrom or so. That isn't good enough to make a clear spectroscopic identification. But way down in the other end of the building resided a big two-meter Bausch & Lomb photographic spectrograph. We wanted to use that. But we couldn't move the laser and we couldn't move the spectrograph.

We gathered a bunch of c-clamps and magnetic bases and put seven mirrors down the hallways, around corners, and through laboratories. We managed to move the laser beam clear down to the other lab, probably three or four hundred feet away. While we were doing this, we were trying to keep it quiet. After all, we hadn't published any of this yet. When you have visitors constantly in a research laboratory, it's very difficult to explain what all these mirrors are doing in the hallways.

We usually operated at night. That was also a little safer, in case one of the mirrors drooped and we scattered laser light all around the halls. We ended up taking some very good plates and making measurements to within a few hundredths of an angstrom that really nailed down the identifications. I am glad we did that because one of the lines had a rather illogical identification. There was another quite logical, rational line (that doesn't lase) that we could have identified it as. The precise measurement paid off.

We got our paper into print in *Applied Physics Letters* in April, 1964. As you may know, there were two other independent discoveries of the same argon ion laser. One was by Guy Convert at CSF in France and the other by Bill Bennett and his graduate students at Yale University. Convert and his associates published in *Compt. Rend.* and Bill Bennett published in *Applied Physics Letters,* but somewhat later than we did. So it was obviously a laser whose time had come.

Q: Did you then go on to the other noble gases?

Bridges: Even before we got the publication written up and off, we started looking at other things. We had other gases on the station, so we went through krypton, xenon, neon, and some mixtures. Lines just sort of tumbled out all over the place.

Neon didn't work initially, and that's reasonable, because the analogous lines were all in the ultraviolet. We didn't have ultraviolet optics covering that range. We ordered some, but before we were able to get anything going in neon, Roy Paananen at Raytheon published his work with ultraviolet lines in neon.

Art Chester, who was a Howard Hughes doctoral fellow at Caltech in the physics department, spent a day a week at Malibu and was assigned to our group. (He is now [in 1990] director of the

Hughes Research Laboratories.) When we started getting this real plethora of lines coming out at us, Art was practically a commuter between Malibu and the Caltech library, where he was photocopying old spectroscopic literature. Unfortunately, noble gas literature is *really* old. Those are the easiest discharges to make, so they were done first. The literature was mostly from the 1920s and 1930s, with more modern, very precise spectroscopy done on other materials. No one went back to redo the noble gases. As we improved our laser setup and went to higher current pulses, we were quickly producing lines that just didn't seem to be in the well-identified literature at all. Then it really became a puzzle to identify lines. Some of the lines we saw with xenon are still unidentified but are probably triply ionized xenon.

A few months later, in June 1964, I was in Washington and went over to the National Bureau of Standards to get some help on these spectroscopic identifications. There I met Charlotte Moore Sitterly for the first time. She presided over the collecting and cataloging of spectral lines and is the author of the famous volumes entitled *Atomic Energy Levels.* She indicated she wasn't aware of any publications I hadn't looked at, but she suggested I might be interested in Curtis Humphreys' unpublished work. Of course, I had already seen his publications from the 1930s. I said, "Oh, are his notebooks still around here? Maybe there was something in his notebooks."

She said, "No, he took them with him."

"Oh, that's too bad," I replied.

She said, "Oh, he's still alive. Why don't you go see him? He is at the Naval Weapons Center in Corona, Calif."

Later, I called him and Curtis was kind enough to send me his ledger papers, hand written in ink, listing his laboratory observations. Happily, we found most of the funny xenon lines in that listing. Now, he hadn't identified them either. But at least it was an independent confirmation that somebody had seen the same lines 30 years earlier, so we weren't crazy.

Art Chester and I put together a publication summarizing all the spectroscopy that we did in this interval. We had a little note added in proof saying that we found this unpublished list of lines that suggested we were probably looking at xenon and not something else. Now, "not something else" wasn't merely an academic question. The tubes we put together used hot thermionic cathodes made of barium oxide. We encountered several other lines we had trouble identifying at first, but after the pattern fit into place, we identified them as ionized oxygen. Our discharge was partly disintegrating the cathode as it ran, contaminating the tube. We also found nitrogen and carbon ion laser lines from contamination and some other things we still haven't really identified.

Q: How did continuous-wave operation come about?

Bridges: A few weeks before the first paper was published, I went to a committee meeting, in New York I think, along with E.I. (Gene) Gordon of Bell Labs. I had told him of the mercury laser work sometime previously. At this meeting I gave him a quick rundown of the argon work and preprints of the argon paper and a krypton-xenon paper that came out sometime later. I told him about the argon pulse being coincident with the current pulse unlike mercury, so that it might be a cw laser if only it didn't chew up so much power.

About two weeks later, while Art and I were deeply involved in spectroscopy, Gene called me and told me, "We've got ours going continuous-wave." That surprised me. Gene had taken the leap of faith that it could be made to run continuous-wave. He took the approach of using a very tiny capillary discharge, about a millimeter in diameter, whereas our tubes were five or six millimeters in diameter. So his current density was up by a factor of 25 over ours for the same current. He also had a superb craftsman making mirrors for him at Bell Labs.

That resulted in a very interesting publication which a lot of people have questioned me on. Gene asked if I would be a co-author on their paper. I thought I ought to do something, since my entire contribution to that point had been to give him the preprints. So I offered to build a little different tube and use our bigger power supply to try to push on to higher power.

I started on that while Gene and Ed Labuda made the first draft of their work. They sent the draft to me and within a week I had a water-cooled tube going with something like 80 milliwatts output. I also got ours running cw with krypton and xenon.

That's how a joint publication between Hughes and Bell Labs

Figure 11.3 The Hughes ion-laser section in 1969, with an assortment of early ion laser experiments. From left to right, Peter O. Clark (holding a *very* early He-Ne laser), Dorothy LaPierre, Donald C. Forster (holding an early metal-ceramic ion laser he had designed), Susan Watkins, Michael Barnoski, Diane Orchard, Heintz Tiergartner, Robert B. Hodge, G. Nield Mercer, Howard R. Friedrich, Ronald Smith, and Bridges (holding a small pulsed ion laser built for the first public demonstration of an ion laser) (courtesy of Hughes Research Laboratories).

came about. It was really just a matter of personal contact. I don't know how happy our supervisors were, but we did it anyway.

Q: How long did you stick with ion lasers and the spectroscopy?

Bridges: I stuck with ion lasers for about six years; the spectroscopy was an on-and-off thing. You have to understand the environment of the times and the environment of Hughes. We had a cw ion laser with almost 100 milliwatts of output just as the first publication hit the streets in April, 1964. The Hughes systems people in

Culver City got on us. They said they had looked at a system about a year earlier and decided that it just wasn't going to be practical with perhaps 50 milliwatts of helium-neon output. But now that we had about 100 milliwatts of green light, they asked if we could give them maybe a quarter of a watt, with some development, for this airborne system. Since we were looking at a laser whose power output went up with the square of the current and didn't seem to stop, it was an easy extrapolation to a quarter of a watt. Too easy, as it turned out.

So, starting in May, 1964, when everybody else was jumping on spectroscopy, we started trying to build a demonstration tube that would give us a quarter of a watt cw and be packageable in an airborne system. The systems people, for their part, started putting the system together.

Through the summer and fall of 1964, we continued to do some spectroscopy evenings and weekends. But our days were mostly spent beating on the laser, doing very practical nitty-gritty development things. Success came fairly quickly. We had a quarter of a watt in a month or so and half a watt not too long thereafter. That was just as well, because the systems people kept coming back and asking for another factor of two. By late 1964, they wanted a two-watt airborne ion laser, so that's what we were shooting for.

The system, to use the *Aviation Week* description, was a scanning night-reconnaissance system. It had a spinning mirror that scanned one dimension on the ground, while the aircraft motion provided the other dimension on the ground. You ended up making a strip map of what's on the ground. Perkin-Elmer had pioneered this system and already had one flying with a helium-neon laser in it.

Other applications came and went over the next several years, but our motivating force was a practical airborne ion laser. The first flight test was in 1965; by 1970 the lasers and the flight tests had become quite sophisticated. But by about 1972 the Air Force, for their own reasons, decided they weren't interested in that particular system. It was expensive, and it worked well. But it had some interesting ramifications on cooling in an airplane. Perhaps that was what made the Air Force a little less than enthusiastic about the system. But when that system interest ended, I think Hughes' interest in argon ion lasers ended, too.

Q: Who were the key people in argon lasers at Hughes during this time?

Bridges: In the 1965 to 1970 time frame, the real players at Hughes were myself, Steve Halsted, Howard Friedrich, Peter Clark, and Nield Mercer. Steve came to us from Stanford University and later, about 1969, went to the Electron Dynamics Division as a department manager and took the argon ion laser into commercial production at Hughes. Steve brought what was probably the nicest low-price argon ion laser to production at that time, the old 3066H. But when that ran its course, Hughes got out of the commercial ion laser business as well.

Nield Mercer had been one of Bill Bennett's graduate students at Yale during the discovery phase. After Nield finished his thesis on ion lasers, he came to work for Hughes. Nield and I spent a bit more than a year working on an ultraviolet version of the argon laser, starting in about 1967. We were able to carry our spectroscopy into the ultraviolet and attain cw operation at new wavelengths. After about a year's work with tungsten disk-bore tubes, we were able to deliver a tube to Fort Monmouth NJ with a 2-watt output in the ultraviolet. Having that kind of laser available allowed us to push some of the lines that had only been pulsed before into continuous-wave mode and reach some higher ionization states.

Q: Considering the state of ion laser development today, are you surprised by any of the commercialization improvements, such as the 25 milliwatt air-cooled versions available these days?

Bridges: No, I don't think there's anything really surprising in any of these lasers in terms of fundamental breakthroughs. They all represent very nice, careful, mature engineering. One thing that often surprises me is that some of the problems we faced in the early days are still around.

If I go back through my files, I will find a rather lengthy proposal we wrote about 1968 to study the window degradation problem. The government wasn't interested enough to fund it then, and commercial companies weren't really involved at that point. But the problems of window film formation and solarization are still around. Everybody has solved the window problem at least ten times over, only to have it come back in one or another form. It's one of those perpetual problems. I'm sort of surprised it hasn't been solved by now.

Q: How about the emerging tube-refurbishment market, where a processor repairs and refills argon laser tubes?

Bridges: Well, we did it all the time. When tubes would fail for one reason or another, we would rebuild them ourselves over and over again as research vehicles. In fact, during our ultraviolet development work, we had one carcass with cathode and anode that had all kinds of different disk structures put in it. It was always a challenge to see how many times you could let a cathode down to air and still have the thing work.

Occasionally, I get a question from another faculty member here at Caltech when they look at the price of a replacement tube. They say, "But it just looks like the windows are dirty. Can't we just pop them off and clean them? Why should I pay $12,000 for a new ultraviolet tube?"

Invariably, my advice is for them to bite the bullet and buy the new tube. If they only knew what was involved in reprocessing one of these big tubes, they would never undertake it on a one-shot basis. Even if they count graduate labor as free, it won't work out. I tell them, "You don't want to go into the ion laser research business. You are a chemist (or an aeronautical engineer). Just buy the tube." But in a commercial sense, it's probably a different matter entirely.

Q: What did you go into after the ion laser effort wound down?

Bridges: I had become the department manager of what was then called the Laser department at Hughes in 1969. If you will recall, there was something called a recession in progress then, so it wasn't a good time to break into management. After about a year and a half of layoffs and transfers in and out, I decided I was too young and tender to do this kind of thing. I really wanted to get back into the laboratory environment, so I resigned as department manager. Peter Clark, who had been my second in command and who had been managing most of the high-power gas-dynamic CO_2 and chemical HF laser activity, took over the department.

I was on the director's staff for a while after that. I found it was very difficult being on the director's staff. The director runs the whole place but doesn't own any lab space. So I had to mooch lab space from the departments.

After a while, I got involved in space communications. The real spark plug there was Frank Goodwin, who, by force of personality, kept laser communications alive through some pretty grim days. I developed an interest in the area when I was department manager. In fact, Goodwin, Don Forster, and I put together an analytical comparison of alternatives for laser space communications, in which we attempted to show the superiority of a coherent CO_2 system over an incoherent Nd:YAG system. The paper appeared in the February 1972 issue of the *IEEE Journal of Quantum Electronics*. Given that no one has put either system up yet, it is still an open question.

Q: Wouldn't that require some significant tuning capacity on the part of the CO_2 laser?

Bridges: For a heterodyne system, you have to track out the Doppler shift from a moving satellite. That meant you had to have a tunable CO_2 laser with perhaps a gigahertz of tuning range. But a CO_2 laser isn't tunable to a gigahertz in its normal form. Part of my time on the director's staff was spent trying to work out some solution.

In mid 1971 Peter Smith at Bell Laboratories published a short paper on the helium-neon laser. The capillary bore of his laser was only about half a millimeter in diameter and long enough that the bore acted as a waveguide. Don Forster came back from the meeting where Smith had presented his paper and said that this might be the thing for CO_2 as well. Incidently, Marcatili at Bell Labs had proposed this structure in 1964; it's amazing that no one built one for seven years. (Actually, Steffen and Kneubuhl had unintentionally built a far-infrared waveguide laser in 1967 or so.)

In any case, I got a project going again with bootleg laboratory space and a borrowed technician to try to build a CO_2 waveguide laser. But I had to put it on the back burner for a while a few months later to assist a classified systems program that needed an injection of instant help.

By the time I got back to the waveguide laser in early 1972, Dick Abrams had come on board from Bell Labs. Peter Clark suggested that he collaborate with me on the waveguide laser. We spent some time on that, from which the first paper appeared in 1973. But I was also involved in other things, since I was on the director's staff. It looked like a good thing for Dick to do, so he took off with the waveguide work.

Then I was reassigned again, probably because I said something was ridiculous and impossible. The powers that be said, "Good, you're just the guy to work on it." That was the adaptive optics work that Hughes was trying to break into. At that time, Rockwell Autonetics had the only government contract, working with a ponderous CO_2 system. Tom O'Meara, our theoretician at Hughes, had some ideas for a much simpler system. Frank Goodwin and Tom Nussmeier had been working on it, but they were reassigned. So making a working version of O'Meara's multidither prescription, as it was called, became my task for 1972.

Later, we got a very nice 18-element system going, using a visible-wavelength laser. That allowed us to make a movie of the system in operation, which is the only way to show off such a dynamic system.

By about mid 1973, the work had grown, and we had hired more staff. Jim Pearson joined us fresh from his Ph.D at Caltech and took over the lab work; I found myself managing again, not only the adaptive optics work, but also another project in "space object imaging."

Q: How did you come to Caltech?

Bridges: In 1974, I had the opportunity to come to Caltech as a Sherman Fairchild Distinguished Scholar. So I did, with a bit of malice of forethought. I figured if I left Hughes for a year, they would have to find other people to manage these programs. Then when I came back I could become a worker again, because the managers would already be in place. It never occurred to me that they would find a manager and then have the manager split the scene about the time I was ready to come back.

I had a great deal of fun teaching a class on optics that year and even did some fiddling around with ion lasers again. I still wasn't able to identify those xenon lines.

I went back to Hughes with a dandy idea for laser isotope separation. I was even able to persuade Hughes management to let me do it. But the price extracted for that was to get involved in another area—the beginnings of the space-borne hydrogen maser clock. That was Harry Wang's project; he was the maser man and my contribution was to work on the atomic hydrogen source and the vacuum system. It was quite an engineering problem: How do you seal this device in a can and put it in a space craft to work for ten years, when it has a bad history of requiring a handful of Ph.D.'s to run it?

Between that and preparing for the isotope separation project, I spent a quick two years. I must confess that the isotope separation project never really got off the ground, because delivery of a tunable dye laser was delayed about 14 months. But in 1977 I had the opportunity to come to Caltech as a full-time professor. I'm not doing much in the way of laser work anymore. I'm spending most of my time on millimeter-wave technology . . . particularly some novel dielectric waveguide components. You might call it a sort of integrated optics for the millimeter-wave region.

But ion lasers are still fun. I like the engineering side of it as well as the scientific side, so I hope to continue on with ion lasers. I have done some consulting for Spectra-Physics, so I still have my hand in it. I may even set up one in the lab here for occupational therapy, to take another shot at those unidentified xenon lines.

Q: What has surprised you most about laser development?

Bridges: There have been many surprises over the past three decades, but the most surprising thing to me is that the argon ion laser is still with us! And sales are at an all-time high. I never imagined the argon-ion laser would last so long.

In the mid-1960s, new lasers were being discovered right and left, and we fully expected that any day something better would come along and take over the "blue laser" business. How hard could it be to find something more practical than a fragile vacuum-tube-like monster that used 25 kW of three-phase electricity and

Figure 11.4 Bridges in his Caltech lab in 1988, with a single-frequency dye laser being pumped by an argon-ion laser (photo by Geraint Smith, courtesy of William B. Bridges).

several gallons per minute of cooling water just to produce a few watts of blue light? An air force general told us in 1965, I think, that we were wasting our time developing the airborne ion laser, since he had just funded some work that would produce watts of blue light from a diode laser at 50% efficiency. And the climate of discovery was so intense then that we believed him. (We didn't stop our work, since Hughes had a flight-test schedule to meet, and that was a fortunate thing indeed, since the "blue diode laser" that the general funded was the "silicon carbide mistake" of the mid-1960s.)

Nevertheless, it amazes me that argon is still a big player, power cable, hoses, and all. It is the workhorse in every scientific laboratory, pumping tunable dye lasers. The more benign (even though less efficient) air-cooled version finds its way into all kinds of graphics and bio-analytic equipment. Too bad my patent filed in 1964 [U.S. Patent 3,395,364] wasn't fortunate enough to be delayed by litigation to a 17-year period when sales were so good. It expired in 1985! Perhaps doubled, diode-pumped YAG or some related solid-state laser will take the business away in the future . . . only time will tell.

Another fact that amazes me is that in 1990 we still await the first demonstration of laser communication in space. I was directly involved in the early Hughes developments of CO_2 lasers for space communications. We proposed a modest bandwidth demonstration (around 100 megahertz as I recall) to NASA in 1968. Frank Goodwin led this quest at Hughes, supported by Tom Nussmeier, Jim Kiefer, and the technicians in their group. Frank and his crew had made and demonstrated several different complete heterodyne systems, first with helium-neon lasers at 3.39 μm, then with small sealed-off CO_2 lasers. We were ready for space, but we never got the chance. The system demonstration for the NASA ATS-F satellite in 1968 went to another company, who failed to demonstrate even a laboratory breadboard, and the ATS-G demonstration was never funded. The air force refused to consider CO_2 for its Project 405B demonstration and went with a YAG-based system. (After many years and about $50 million, the air force abandoned the effort.) My bookshelves at Hughes were overloaded with proposals. Here we are in 1990, with Lincoln Labs reporting demonstration of all the pieces of a gallium-arsenide-based system, but still waiting for someone to authorize a space shot. I can imagine *their* frustration, since I know they've been working on it for at least 20 years.

Q: Where do you think laser technology is going?

Bridges: I'm probably the worst person in the world to ask. I'm the guy who was really *worried* that a silicon-carbide diode would kill off argon in 1965. I should also say that when a couple of Hughes staff members in 1966 told me that "home computers" would be the big thing in the future, even bigger than amateur radio (the three of us were hams), my reply was "You've got to be kidding. Who would spend money to play around with ping-pong games when they could communicate all over the world by ham radio?" I sort of missed that one!

One thing I have observed over the years, however, is that the laser often plays the role of a catalyst. Optical fibers were known long before the laser was even conceived, yet they did not receive wide attention and application. (I recall being fascinated as a 6-year-old by a plastic light pipe used to direct light down my throat during a medical exam.) The laser came along and stimulated the development of low-loss fibers. Now all kinds of fiber applications are being considered that don't use lasers . . . its catalysis is accomplished. I would expect in the future to see a much wider application of nonlaser optical systems.

Q: What are you doing today?

Bridges: I am working with students at Caltech in 1990 in three different areas. First, we're working on very-high-frequency electro-optic modulation. We have an idea to solve the phase velocity mismatch problem in lithium niobate waveguide modulators by coupling the modulating microwaves (or millimeter waves) into the electrodes by using antennas on the $LiNbO_3$ substrate. We have just demonstrated an 8–12 GHz model, and are beginning to fabricate a 60-GHz version. This should allow very wide band subcarrier analog modulation in fiber-optic systems.

Second, we are exploring ways to improve CO_2 waveguide lasers (a return to the work I started at Hughes in 1971). We've successfully made coupled waveguide laser arrays and open "slab" CO_2 lasers, and we're now exploring better waveguide materials and better resonator configurations. We want to generate a lot of 10-μm power efficiently from a very small package.

The third area is dielectric waveguides for millimeter waves. I started about 10 years ago, and we demonstrated a small, flexible guide with loss of 5 decibels per meter at 100 GHz. It's not much greater than a conventional metal waveguide, but it isn't exactly 1 dB/km. Since then, we've worked on other dielectric waveguide components: directional couplers, ring resonators, and so on. Our hardware looks just like the pictures in Stew Miller's famous 1969 *Bell System Technical Journal* article on "integrated optics," except that our components have millimeter dimensions where his had micrometer dimensions.

Of course, my curiosity often gets the better of me, and I undertake quests not "practical" enough for funding agencies. We academics are supposed to be both "blue sky" and "relevant" at the same time, but I'm more often "just curious." I usually take these fliers with undergraduate students because Caltech undergraduates are more fearless than the graduate students. They have to be much quicker, since they graduate so soon, and they work for free. Over the last few years, we've looked at things like techniques of optogalvanic spectroscopy, measuring microwave photodetectors by photomixing tunable lasers, making artificial "nonlinear materials" in the microwave range by using arrays of semiconductor diodes stuck in dielectric waveguides, trying to tie up some loose ends in the "thermal blooming" of light passing through liquids, and trying to understand some puzzling frequency-pulling effects in He-Ne lasers.

Q: If you had it to do over again, would you still get involved with lasers?

Bridges: You bet! Those were exciting times, the kind of thing that happens only once in a lifetime. I wouldn't have missed them for anything. And I feel the same way about these last years at Caltech, too.

An earlier version was published in *Lasers and Optronics* ®
(formerly Lasers and Applications) a Gordon Publications, Inc. publication.

WILLIAM T. SILFVAST

Metal-Vapor Lasers

For sheer number of new lasers discovered, it's hard to match the record of William T. Silfvast. His doctoral research included demonstrating laser action for the first time in vapors of nine elements. Much of his work has involved metal vapors; the most tangible commercial result is the helium-cadmium laser. He also has demonstrated over 100 recombination lasers, laser action in laser-produced plasmas, and high-gain photoionization lasers pumped by soft x rays. His most recent work is on short-wavelength lasers.

A native of Utah, he received two bachelor's degrees, in physics and in mathematics, from the University of Utah in 1961, and received a doctorate in physics from the school in 1965. During a year of postdoctoral research at Utah, he discovered the 441.6-nanometer blue HeCd laser.

He spent the following year at the Clarendon Laboratory of Oxford University under a NATO Postdoctoral Fellowship, then joined the technical staff at AT&T Bell Laboratories in Holmdel, N.J., in 1967. He remained at Bell Labs until December 1989,

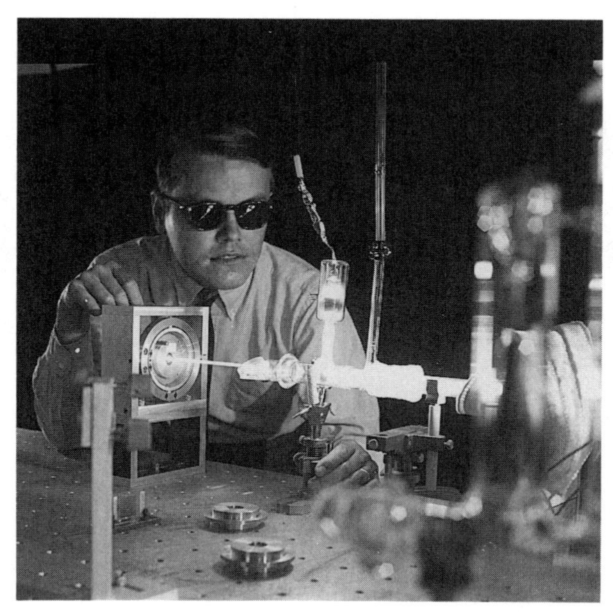

Figure 12.1 William T. Silfvast aligns the discharge tube of a metal-vapor laser at Bell Labs in 1968 (courtesy of AT&T Bell Labs).

except for a 1982–1983 sabbatical year at Stanford University, where he worked on short-wavelength lasers under a Guggenheim Fellowship. He was named a Distinguished Member of the Technical Staff at Bell Labs in 1983. In January 1990 he was named a professor of physics and electrical engineering at the Center for Research on Electro-Optics and Lasers (CREOL) at the University of Central Florida in Orlando.

He is the author of nearly 100 technical papers in reviewed journals and holds 18 patents. He has served as program and conference co-chair of CLEO (the Conference on Lasers and Electro-Optics), has been a director of the Optical Society of America and chair of its technical council and nominating committee, was an associate editor of the *IEEE Journal of Quantum Electronics,* and served on a review panel for the National Bureau of Standards. He was elected a fellow of the Optical Society in 1987 and of the Institute of Electrical and Electronics Engineers in 1989.

Jeff Hecht conducted this interview on September 28, 1984, in Silfvast's Bell Labs office. It was updated in 1990.

Q: How did you get involved with science?

Silfvast: From the earliest I can remember, I wanted to be a mechanical engineer. I always liked to tinker. But when I was studying engineering at the University of Utah, I got pretty frustrated, because all I was doing was memorizing formulas. After two and a half years I quit and went to California with a couple of friends looking for a job. I wound up doing engineering work for Lockheed on the Polaris missile project, and after almost two years of that I realized that I had to go back to school and that I didn't want to be an engineer. I decided to go back to Utah and study physics. I enjoyed it so much that I decided to go on to graduate school, even though I hadn't planned to. In my senior year I took an optics course and fell in love with optics. It was just about that time that the laser was discovered, and I became very enthused about lasers as well. I started working with Grant Fowles, who was also getting interested in lasers. Before long, he and Russell Jensen discovered the first charge-transfer laser in iodine vapor.

Q: What was your role in this?

Silfvast: Fowles thought it would be worth trying to make a laser out of bismuth to study hyperfine structure, since it is one of the heaviest of the odd elements and consequently has a large hyperfine splitting. It also has a reasonable vapor pressure. I accepted the challenge and scrounged some zone refining furnaces from the ceramic engineering department as well as other miscellaneous parts and put together a laser discharge system.

There had been a couple of metal vapor lasers, in discharge-excited mercury and optically pumped cesium. I was not aware of the cesium work at the time, and the mercury laser didn't require heating to vaporize the metal. The mercury work, however, no doubt inspired us. We didn't have much financial support, but Fowles was ingenious at putting together inexpensive apparatus, and I adapted quickly to that. We made a quartz tube that we could just attach to simple electrodes and used microscope cover slips mounted with vacuum grease as Brewster angle windows. I think that system was unique then, because I could change from one metal to another in a couple hours and try a lot of experiments in a short time.

Q: That must have been one of the first systematic searches for new lasers.

Silfvast: It turned out that it was, but we didn't start that way because we were focusing on bismuth. We didn't know how long the metal would last and thought we might have to change tubes very often. Also our approach was inexpensive, if you had a good glass blower.

That tube design turned out to be a key element in our success. Later we found out that some people in Gene Gordon's group at Bell Labs were thinking about metal vapor lasers. They apparently were slowed down by building elaborate high-vacuum discharge systems, and we got our results first. Also, we didn't have preconceived notions of how pure a laser tube should be—it had to be very pure in the types of lasers discovered before then—and in some ways our naivete was a help.

Q: What happened with bismuth?

Silfvast: I spent a good three or four months looking for lasers at various wavelengths, mostly in the visible, but never saw a laser at all and was somewhat discouraged. Then one day in early 1965 I said, "Hey, we're not getting anywhere," and thought about mercury. Looking at the periodic table, I saw that cadmium and zinc had similar electronic configurations. So I went to the chemistry department stock room and obtained some purified zinc and cadmium.

I put zinc in the tube first, and the very first time I turned it on I got this turquoise, blue-green transition at 492.4 nm to lase. It was on a Saturday. I was just ecstatic and went running around the campus looking for Fowles. I found him in a committee meeting, which he immediately left and came running down to the lab to see the laser.

We studied that system for a few days, and then we put cadmium into the discharge tube and made many of its transitions lase, although not the blue one yet. The transitions we got to lase were those that seem to work by charge exchange. We decided that many of the metals were the most interesting candidates for lasers and started trying every metal we could find that we could vaporize in our quartz discharge tube, which went up to 1200°C. Besides cadmium and zinc, we made lasers in gallium, germanium, tin, indium, and so on, but the next really exciting one was lead.

A lot of the early gas lasers were discovered somewhat accidentally by researchers looking for another transition. Gas laser excitation mechanisms were too complex to model accurately at that time, though we do a lot better now. We really were trying to make lead ion lasers. In fact, we had made them. But one day, when we had blue mirrors on, I saw what was obviously a red laser beam. It was exciting. I've discovered my share of new lasers, and every one was very exciting because it was something new that you were seeing for the first time. In this case, I said at first that this can't happen—the mirrors must have had losses of 70% to 80% in the red, yet I saw the laser beam. It turned out to be the high-gain self-terminating neutral-atom lead laser that led to the whole series including manganese and copper.

Bill Bennett heard about our work and flew out to see it. We were impressed that he came out; he was one of our heroes. We didn't know that at the time he was consulting with TRG where they were working on Gordon Gould's idea of a collision laser. They had been running a continuous-wave discharge in manganese, and we were later told that after Bennett saw our work he told them to start pulsing their discharge. Later, while we were making manganese lasers, all of a sudden Fowles got a letter on TRG's work in manganese to review for *Applied Physics Letters.* They had scooped us, though I'm sure in the long run it was good to have Bill Bennett visit us.

Q: Were other laser researchers surprised at this little group way out in Utah?

Silfvast: Fowles gave a paper at the Electron Devices Meeting in Urbana. Illinois. Besides zinc, cadmium, and lead, we also had made phosphorous and sulfur lase. We scooped two Bell Labs groups! Colin Webb, Ed Labuda, and Dick Miller approached us to say that they were thinking about making metal vapor lasers. And later Peter Cheo told us he had planned to give a postdeadline paper on phosphorous and sulfur lasers at another meeting in a month or two. I can't tell you how thrilling it was for me because I thought of Bell Labs people as the ultimate leaders in research.

Q: What did you do after that?

Silfvast: I did my thesis on the lead laser. I predicted some incredibly high gains, 6 to 10 decibels per centimeter, much more than anybody else had reported back then and higher than we knew how to measure. Then I stayed on at Utah for another year as a postdoc. One of the things I did that year was study some peculiarities I had seen in the laser characteristics of some of the metals. We had seen bright blue emission, not laser emission, from a couple of zinc and cadmium transitions. I went back and studied them. In 1966 I first observed the 441.6-nm blue cadmium laser and found that it worked only with helium. We hadn't made it lase the first time around because it required very low excitation.

Q: Had you been working with nominally pure metal vapor before?

Silfvast: No, always with a buffer gas to get the discharge out of the heated region to the electrodes. We hadn't realized that helium was essential for some transitions. We were going through so many lasers that it would have taken at least 10 years to work out the details of the excitation mechanisms.

I applied for a NATO Postdoctoral Fellowship to work with John Sanders' group at Oxford University—he was doing some laser work, and I liked the idea of going to Oxford. I got the fellowship, which was the first thing I had ever won in my life, and it was very exciting. One of the things we did at Oxford was to develop a technique for measuring very high gains that showed the lead laser gain was 6 to 7 dB/cm, agreeing with the calculations I had made earlier.

Meanwhile, Bruce Hopkins, a graduate student at Utah, continued studying the HeCd laser. He saw some quasi-continuous-wave emission in the blue, at the peak of the ac excitation cycle.

When I came back and visited, Fowles showed me the results, and they looked promising for a continuous-wave laser.

Q: You did that work here at Bell Labs. How did you come here?

Silfvast: I interviewed with Bell Labs while I was at Oxford and was very excited when they made me an offer. I was very surprised at the salary, several thousand dollars more than I would have accepted as a minimum to work at Bell. At first I came only for two years. I wanted to teach, and I think Bell had not seen enough of me to risk offering a permanent post. But I liked it so much here that, when they offered a permanent post, I took it. I didn't have strong training in optics as a student, since PhD exams were offered only in solidstate and nuclear physics. It took me a good ten years to get over my feelings of inferiority at Bell, even though things went very well for me.

Q: Did you start right in on metal vapor lasers?

Silfvast: Not exactly. First, Kumar Patel suggested I work on carbon-dioxide lasers, something of more interest at that time than metal vapor lasers. So I started coming up with some ideas there, but meanwhile I talked more with Patel and P.K. Tien, who was my department manager. I convinced them that there were some interesting things in metal vapors. Also, Patel was working on self-induced transparency, which was a hot thing back then, and we thought that the lead laser would be useful for that. I started dividing my time between the lead and cadmium lasers.

I came here in August, 1967, starting with a completely empty lab, and by November or December I had a HeCd laser working continuous wave for the first time. By the following June, I was able to give a postdeadline paper at the International Quantum Electronics Conference in Miami reporting continuous-wave, fairly efficient operation of the 441.6-nm HeCd laser.

Q: What was the key to continuous-wave operation?

Silfvast: Learning how to run a steady dc discharge in a metal vapor such as cadmium, running a low enough excitation current, and getting the proper vapor conditions. At Utah we used ac neon sign transformers that we had scrounged. They already had built-

Figure 12.2 Silfvast and Obert R. Wood II use an electron beam to pump lasers at Sandia National Laboratories in 1974 (courtesy of Sandia Laboratories).

in ballasts and were much easier to operate in controlling the vapor pressure when we were just looking for lasers. Running a dc discharge is harder because you have to worry about contaminants in the discharge and cataphoresis effects. The conditions we needed were quite different from those in the argon laser which had already been discovered by Bill Bridges.

Tom Sosnowsky, here, discovered the cataphoresis effect, and I think John Goldsborough also did at Spectra-Physics at the same time. We had been producing the metal vapor by putting the metal in little cups or dimples in the discharge bore. Then Sosnowsky found that cataphoresis in cadmium is high enough that a single source of metal at one end of the tube can produce a fairly uniform discharge over the entire length of the tube. That was a significant step in the development of a commercial product because it made the tube design simpler.

Q: Who discovered the 325.0-nm ultraviolet line of HeCd?

Silfvast: John Goldsborough and I did that independently, but I think his publication was first. The discovery of cataphoresis helped make a more uniform gain region which helped overcome both the mirror problems at 325.0 nm and the reduced gain on that transition.

Q: How did the HeCd laser become a commercial product?

Silfvast: I think Goldsborough built the first model at Spectra-Physics in 1969 or 1970. I inherited one of those old monsters from somewhere, and it's still lasing. It was a couple of meters long, very expensive at the time, and it was easy to see that it would never be practical. RCA came out with a similar laser not long afterwards.

About that time, a fellow from a development area here came to me with a new idea for a remote blackboard. The idea was to transmit in real time only the changes in the chalk motion on a blackboard during a lecture. This would take only the bandwidth of a single telephone line. If you were scanning the entire blackboard for changes, you'd need the bandwidth of a television channel.

They found that the HeCd's ultraviolet line could write on a DuPont product called DyLux and wanted to use laser writing and projection of the image for a display. They came to me looking for a small, inexpensive ultraviolet HeCd laser. I had some ideas about what we later called the segmented-bore HeCd laser. If they worked I could build a HeCd laser that would fit into a suitcase, so I started on the project.

We put little cadmium washers between glass discharge tube segments and demonstrated 1,000 hours of operation in this laser, which was self-heated by the discharge. It gave 15 milliwatts in the blue, as much power as the RCA model that was three or four times as long, and it also produced 2 milliwatts in the ultraviolet. No one had ever made a laser that short work in the ultraviolet before. And it used the natural mixture of cadmium isotopes instead of the costly single isotope.

I was doing this all on my own, and with that success I was sort of committed. At one point I mentioned to management that these development people wanted a laser, and they said, "We don't want to be a laser supplier, that's not our purpose here." But I kept working on the project. That's the freedom that we have here at Bell Labs.

After I had demonstrated the laser, the remote blackboard people got very excited. The laser was good enough to let us think about a commercial product. So I sent descriptions of my work to five companies, looking for two of them to build three or four development models each for the remote blackboard system. Some companies said it couldn't be done, but by the time we got out to Coherent Radiation (now Coherent Inc.), they already had one working. They had some good ideas, so our development department gave them a contract and also gave one to Hughes, which had other good ideas.

Jim Hobart at Coherent liked the laser and put Mark Dowley on the project. Mark worked on it for about two months and decided it was such a good idea that he would start his own company. So he left Coherent and started Liconix.

Q: Whatever happened to the remote blackboard?

Silfvast: The ultraviolet laser was not far enough along in development, and they found out they could use a television display instead—it's a commercial product now. Tony Berg, who was then at Bell, also built a very competitive scanner and microfilm printer that used the blue HeCd laser. However, even though it was a good system, it never made it to the marketplace—it wasn't the kind of thing that AT&T was geared to sell in those days.

Q: Has anything held back commercial development of HeCd?

Silfvast: There were some problems in the early days with the company that advertised an inexpensive HeCd laser before their laser was proven. They sold many lasers at a very low price. When their laser suffered problems and was withdrawn, this gave the laser a bad name for a while. But the real problem has been a lack of investment in the technology to bring prices down. The HeCd discharge tube can be made almost as inexpensively as a helium-neon laser. The power supplies do have to be more expensive than HeNe, and because you have to heat the discharge, it is difficult to use internal mirrors. Both Liconix and Omnichrome have some very good lasers for sale now with 5,000 hours or more of operating life, but the real low-cost laser has not yet happened.

Q: Why has HeCd been the most successful metal-vapor laser?

Silfvast: Because it is simple and has the right wavelengths. Cadmium happens to have the right electronic configuration. By just removing an inner shell *d* electron from the ground-state configuration of cadmium, Penning ionization from helium metastables or direct electron or photon ionization of cadmium puts you in the upper laser level of the ion. This is in contrast to the normal situation where removing a single electron just puts you in the ion's ground state, which of course can't be inverted with respect to other ion states. Similar levels exist in zinc, but the transitions are in the red and yellow, which are not as interesting, and zinc needs a higher operating temperature.

Q: Whatever happened to helium-selenium? You had that on the cover of the February 1973 *Scientific American*, and there were some commercial models in the mid-1970s.

Silfvast: I think HeSe can simultaneously put out more wavelengths than any other laser. We've been able to lase about 25 lines from the blue to the red at the same time. When Marv Klein and I first discovered it, there was no continuous-wave dye laser, and we thought it would be a big commercial success. RCA and Liconix offered commercial models, but they had to pull them off the market because selenium vapor pressure is much harder to control than cadmium—it tended to wander around and end up in the wrong places. I believe those problems could be solved, but there hasn't been a strong application need to justify that investment.

The cover photo, which has become relatively well known, was an interesting story. Fritz Goro, the *Life* photographer, got the credit, but Leo Szeto and I actually devised the way to take the picture, which shows the many laser lines spread out in multiple orders from a diffraction grating. Smoke didn't work, so we finally painted the beams onto the film by moving a large white card back and forth.

Q: What did you do next?

Silfvast: I moved from metal vapor lasers to recombination lasers, where excitation comes from recombination of a free electron with a positive ion-excitation from above. Obert Wood had made one of the first TEA carbon-dioxide lasers, which produced bright plasmas by focusing the pulses in the air. I thought we ought to be able to make a laser in such a laser-produced plasma. Eventually, in 1976, we succeeded in making the first laser-produced plasma laser, using recombination in rare gases.

Later Obert and I developed probably the simplest laser ever discovered, the SPER laser, for Segmented Plasma Excitation and Recombination. It consists of a series of electrodes—as few as two—that you can apply with double-sided tape, a couple of mirrors, and a low-pressure gas in the cavity. With proper electrode spacing, you can produce a metal vapor spark in the electrode gap region rather than a gas discharge, using either high voltage or—we found out later—low voltage. The high-density metal vapor spark plasma expands, and laser action in the recombination spectrum occurs in the periphery.

We've discovered over 100 new laser lines this way. It works with a lot of metals that we couldn't easily vaporize in other laser designs. Some designs are very easy to make. We call them BIC lasers because we think they could be sold for a couple of dollars if you made them in large quantities. They produce peak powers of a few watts in pulses 10 to 100 microseconds long, and we've had some ideas for applications. The wavelengths range from just under 300.0 nm out to 5 micrometers. Most are in the neutral species in the infrared. The visible ones are mostly in the single ion, and the ultraviolet ones mostly in doubly-charged ions, because energy levels get farther apart with higher ionization levels. We had hoped to go into the vacuum ultraviolet, but the SPER lasers can't produce significant excitation beyond the double ion. So we set that work aside, although it has interesting commercial applications, because our real goal was short-wavelength lasers.

Q: Have you found a more promising approach to short wavelengths?

Silfvast: Obert and I did some more work with laser-produced plasmas. Then in 1982 I received a Guggenheim Fellowship and spent a year with Steve Harris at Stanford. It was a very stimulating time, during which I got the idea of trying inner-shell photoionization using soft x-rays to directly pump cadmium. The equipment at Stanford was tied up with other experiments, but I did the modeling out there, and we set up the experiment back here. It was one of those unusual times where I had the chance to do all the theory first and consequently knew the exact experimental conditions to set up. It worked on the very first shot.

We focused a high-power laser onto a target in a heat pipe containing cadmium. That produced a plasma that emits x-rays, which pump the cadmium vapor to produce the particular *d*-electron state of the 441.6-nm blue transition, giving very strong inversions. Mike Duguay had predicted that these inner-shell transitions could be used for an x-ray laser back in 1967, but no one had ever made a laser this way until we did in 1983.

Q: What are you working on now?

Silfvast: My work on short-wavelength lasers got me wondering what to do with such lasers. Obert Wood and I began thinking about lithographic applications. In 1986, we proposed that the Department of Energy investigate using a soft x-ray laser as a source for projection lithography to make microchips with 0.1-μm features. The short wavelength would overcome diffraction problems in imaging such fine features. At that time, two processes were being developed for thin line lithography at high production rates—excimer-laser projection lithography and x-ray proximity printing—but both were limited to 0.2 to 0.3 μm features. There also was the electron-beam direct-write process, but it was a serial rather than a parallel process, making it too slow for mass production.

Our x-ray laser source received high marks, but the reviewer said soft x-ray projection lithography could not be done in a production setting. The argument was that there were no suitable resists or large field diffraction-limited imaging optics, and that mask alignment was beyond present capabilities. Saying that it couldn't be done was enough to spur us on. We convinced Bell Labs it was a worthwhile project, and we were off and running. Eventually we teamed with others at AT&T, including Rick Freeman, John Bjorkholm, Don White, Tanya Jewell, and Don Tennant, and demonstrated 0.05-μm features imaged with a 20-power reduction optic using 130-angstrom radiation from the Brookhaven synchrotron.

This effort spurred my interest in soft x-ray imaging and the further use of laser-produced plasma sources. Since joining the faculty at the Center for Research on Electro-Optics and Lasers (CREOL) at the University of Central Florida, Martin Richardson and I have established what we are calling a laser plasma laboratory for science and technology to exploit all aspects of laser-produced plasma x-ray and soft x-ray sources, including lithography, microscopy, surface science, atomic physics, and so on. I'm excited about my new career as a faculty member, and, besides continuing to do research and working with students, I am particularly excited about helping maintain and/or regain our country's edge in laser and electro-optic technology.

Q: What has most surprised or impressed you about laser development so far?

Silfvast: I am most surprised at the applications that are using the most lasers. In the 1970s, the most lasers were used in sewer pipe alignment. In the 1980s it was compact disc players. It is interesting to speculate what it will be in the 1990s. Perhaps it will be a monolithic device, including a diode-pumped neodymium laser operating multifrequency and doubled to produce red, green, and blue lasers on a single chip. Such a chip could be installed in every telephone receiver to project a color video display onto a nearby wall or a small screen. These applications are quite different from the anticipated major use of lasers, for long-distance communications. The continual reduction in the losses of optical fibers has reduced the number of lasers required for that application.

A second surprise is the development of optical fibers as a transmission medium. When I first joined Bell Labs in 1967, Rudy Kompfner was studying rain patterns to determine the feasibility

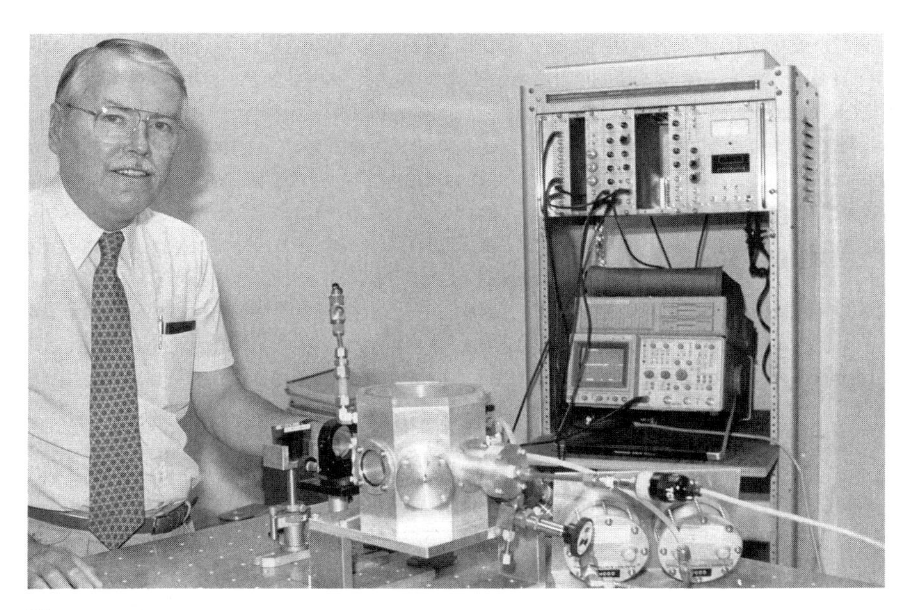

Figure 12.3 Silfvast in his laboratory at CREOL in Florida in 1990 (courtesy of W. T. Silfvast).

of laser communication through the atmosphere. No one then considered optical fibers because of their high losses. In hindsight, I was quite fortunate to have been in the midst of the early development of fiber optics and integrated optics even though I was not directly involved in any of that work. I was sort of a maverick at Bell Labs in that sense. I was working on HeCd lasers, laser-produced plasma lasers, recombination lasers, photoionization lasers, and soft x-ray lasers, all of which would probably have no significant impact upon the business of AT&T. But what a strength AT&T had during those years. I could work on anything I wanted and only had to write three one-paragraph progress reports a year. Such freedom was certainly responsible for developments at Bell Labs such as the transistor and the laser by providing the climate to explore one's own curiosity instead of being pressured to work on someone else's problems.

Q: You mentioned a multicolored visible laser on a chip. What other new directions do you expect laser technology to take in coming years?

Silfvast: In a broader sense, the direction will be toward domination by solid-state laser devices, which is already happening in 1990 and will accelerate. The fundamental reasons are size, operating lifetime (or reliability), and cost. Every user prefers a smaller laser. A solid-state laser has an inherently higher density of laser species packed into the amplifying medium than do gas lasers, approximately 1,000 to 1,000,000 times more for a given amplifier volume, thereby significantly shrinking the size. Reliability is better because the laser species in a solid can't move around and react with potentially destructive contaminants, as it can in a gas or liquid. We don't even have that kind of reliability with ordinary light bulbs.

This long lifetime of solid-state lasers (including semiconductor lasers) is possible largely because of better growth techniques for the solid-state media. Improved purity, damage-resistant materials, and the advent of unique structures and quantum wells have been the driving forces of the solid-state laser revolution.

Inexpensive green and blue semiconductor lasers will most certainly appear on the market in the next decade. Someone will figure out how to make zinc selenide and zinc sulfide materials lase in a small electrically pumped device operating at room temperature. That will put pressure on the HeCd and argon laser manufacturers. But the helium-cadmium laser has played an important role in the laser industry, and continues to do so today, and I have no complaints about that.

Q: If you had to do it over again, would you do the same kind of work?

Silfvast: It is so tremendously exciting to make a new laser that I can't ever imagine not looking forward to coming to work every day, even in slow times. I've probably made more new types of lasers than anybody else, if not in numbers of wavelengths at least in different species. Many of those new lasers have opened up new areas of research and development, and that's the kind of excitement that keeps me coming back for more!

An earlier version was published in *Lasers and Optronics* ®
(formerly Lasers and Applications) a Gordon Publications, Inc. publication.

JAMES J. EWING

Excimer Lasers

James J. Ewing received a doctorate in physical chemistry from the University of Chicago in 1969. After serving on the chemistry faculty at the University of Illinois in Urbana and the University of Delaware in Newark, he joined the technical staff of the AVCO Everett Research Laboratory in 1972.

At AVCO Everett he and Charles A. Brau were the first to demonstrate three key excimer lasers: krypton fluoride, xenon fluoride, and xenon chloride. He also worked on laser enrichment of uranium isotopes before moving in 1976 to the Lawrence Livermore National Laboratory, where he was program manager for the "Rapier" advanced excimer fusion laser technology testbed.

In 1979 he was named director of the laser technology group at Mathematical Sciences Northwest in Bellevue, Washington. For several years the company was known as Spectra Technology Inc.; it is now called STI and is a subsidiary of Amoco. Ewing is now senior vice president, responsible for laser programs and products as well as company-wide business development. Laser work at STI includes laser radar and remote sensing (lidar), custom laser

Figure 13.1 Charles Brau (front) and J. J. Ewing conduct early excimer laser experiments at AVCO Everett Research Laboratory (courtesy of Charles Brau and AVCO Research/Textron).

systems, production tunable solid-state lasers, and development of free-electron lasers, undulators, solid-state and nonlinear optical technologies, and gas-laser technology.

C. Breck Hitz conducted this interview on December 13, 1984, in Ewing's office at Spectra Technology Inc. It was updated in March 1991.

Q: What motivated you to look for laser action in excimer molecules in the first place?

Ewing: Beginning around 1972, when I first joined AVCO and started working with Charlie Brau, there was a big push to find efficient visible lasers. There had been success at making efficient 5- and 10-micrometer lasers, carbon monoxide lasers were looking like they could be 40% or 50% efficient, on paper anyway, and carbon dioxide was scalable to high power at over 10% efficiency. And there was work on hydrogen fluoride going on, some of it at AVCO but most of it elsewhere. It looked like one could make good lasers in the infrared, but the question was, "Why can't you make them in the visible?" And excimers were one of several candidates for good visible lasers.

Q: Is there any one person that you can put your finger on as being the first to conceive the idea of an excimer laser?

Ewing: Well, you have to understand, there are two kinds of excimer lasers. People were thinking about pure rare-gas excimers for a long time, since even before I got into it.

Q: You mean rare-gas dimers, like Xe_2 and Kr_2?

Ewing: Right. I'd have to look up who was the first person to come up with that idea. Jim Keck at AVCO had suggested using a dissociative state as a lower laser level many, many moons ago, and he was thinking about hydrogen. Houtermans had the concept of using mercury molecules as a laser medium. Around 1971 or so there was a variety of people who proposed different things in excimers. Charlie Rhodes, then at Livermore, was one of the most prominent, and he really got the first of those experiments on the air working on pure rare gas excimers: xenon, krypton, and argon. There was rare gas dimer work at Sandia (Jim Gerardo and Wayne Johnson), Northrop (with [Mani] Bhaumik, [Earl R.] Ault, and others) and work in the Soviet Union around the 1972 and 1973 timeframe.

Within AVCO Charlie Brau was the first to champion the original suggestion which came from Houtermans. We embarked on mercury-based excimers, but I don't think any one person could be credited as being first to say, "Excimer lasers, that is something to go after."

Q: What about the rare-gas halide excimer? Can you pinpoint where that idea came from?

Ewing: First off, just a peculiar comment on English: "Excimer" is a word from organic molecular physics. It's short for "excited dimers," and "dimers" means two of the same thing. It's definitely a misnomer as we use the word in laser physics to refer to things like KrF. There is a right name for these molecules: "exciplex." But it doesn't rhyme with laser, so if you call it an "exciplex laser," that's a big mouthful. "Excimer laser" sounds better. My own belief is that people just call them that because it sounds better.

But my involvement in the rare-gas halide work came about because of a meeting that I didn't attend. There were two university labs who were working on it ahead of us at AVCO. One was Don Setser, a professor at Kansas State. He had done a lot of work on energy transfer from argon's metastable excited states. He was doing kinetics in low-pressure flowing afterglows. He presented a paper at the 1974 chemical laser meeting in St. Louis. I wasn't there. But Hao-Lin Chen, Charlie Brau, and Bob Center from AVCO were there. They heard his talk, got excited, and came back and said, "Setser gave a paper on how he reacted xenon metastables with halogens and got this spectra which had to be produced by a molecule of xenon and chlorine."

But the real interesting question to me was, why on earth was the spectrum of xenon excimers at 170 nanometers while xenon chloride was at 300 nm, if they're the same kind of molecule? That was the key question, and that was a question that was posed to me by Charlie Brau, I believe on the next Monday.

So that's what originally got me interested in rare-gas halides. There was this basic property of excimers, we'd seen it in mercury excimers, xenon excimers, and things like that. The resonance level is at about 10 electronvolts and the emission is at 8 eV. There was maybe 1 eV of binding energy in the excited state, everyone could understand and believe that. And there was 1 eV of repulsion in the ground state. Ten eV minus two puts the emission at 8 eV, right where it belongs. So why was XeCl emitting around 5 eV? Well, I remember sitting down and thinking that the excited states of the rare gases are very similar to alkali atoms. They have very low ionization potentials. So one of the things that I realized early on was that these new molecules had a nature that was kind of like that of an alkali halide. And I put together a very crude model and discovered that, based on what happens in an alkali halide, it's not surprising at all that XeCl emits around 300 nm.

Next we said, "Let's try an experiment. Let's put some iodine and some xenon in front of our e-beam, zap it, and see where the light comes out." (Xenon iodide, by the way, has never lased, but it was the first thing we tried to make fluoresce.) According to my crude model, the light was supposed to come out somewhere around 250 nm. So we zapped it, and a tremendous amount of light came out.

There were two things we noticed. One was my wavelength prediction was within 20% of where it should have been. So we said, "Maybe there's some veracity to the model." Second, the emission was brighter than anything we'd been working on for the last three years. We said, "What have we been doing wrong all this time?" It was sort of an exciting thing. I remember sitting down and saying, "Well, if this is where it is, what really do those potential curves look like?" I came up with a different model that went into our first paper, which was in *Physical Review*. That was a classic story, in and of itself.

Q: You had problems with journal referees?

Ewing: We sent this paper describing the new model to *Physical Review Letters,* and the referee sent it back saying, "Well, it's nice, but all it is is spectroscopy; you looked at some spectra, so what?" This was a classic case of missing the significance of a paper. So we said, "Send it to another referee." Another referee said, "Well, it's an interesting thought, but maybe you ought to look at a few more molecules to see if that theory is really correct." By this time we said, "Just put it in *Physical Review*." We didn't care if it came out in *Physical Review Letters*.

But it kind of galled me at the time. I remember, it was April Fools Day of 1975 when Charlie Brau put this made-up letter from *Physical Review Letters* on my desk. It said something like, "Well, we've sent your paper out to this reviewer and to that reviewer, and they think you ought to go back and remeasure everything," and so forth and so on. I looked at that; it made me so irate! But he was just pulling my leg.

Q: So it was Setser's observation of fluorescence from XeCl that eventually led to your work with rare-gas halides?

Ewing: That's true. Independently Golde and Thrush in England were looking at argon excited state reactions with chlorine at low pressure. But one of the things that was curious was that the spectrum that we observed was entirely different from what the low pressure work showed, and that had to do with what pressure we were at. Setser did things at very low pressures, and his bandwidth was huge. When you excite the rare-gas halides at high pressures, the emission bandwidth is considerably narrower.

So, if you calculated the gain coefficient based on Setser's data, you might conclude that these molecules aren't very interesting for making lasers; the spectrum is so broad it would never lase anyway. But at AVCO, we looked at Setser's data and asked, "Why is the emission at that wavelength in the first place?" And that was the right question to ask because it led to an understanding of rare-gas halide molecules, and that understanding led to the laser.

Q: When was all this going on?

Ewing: Well, it was in '74 when we started working on this stuff. September or October of '74, and our first paper was ready to be submitted in December of '74. It wasn't until the next June that we actually got one to lase.

Q: What was the first rare-gas halide laser?

Ewing: That was xenon bromide. I guess it was around March of '75, when there was a briefing at Naval Research Labs for the 50 or 60 people working on these projects. There were Phelps and Gallagher from JILA, Charlie Rhodes and the Northrop group working on argon-nitrogen (Ar/N_2) plus all the people from Naval Research Lab, and others. We gave a presentation outlining where all of the excited states of these molecules should be, and where all the emission bands should be, and so forth. We showed the high-pressure spectra of all the xenon halides. Afterwards, Stu Searles of NRL pulled me aside and showed me their work. They started it about the same time, having been excited by Setser's talk.

What they were working on was XeBr and what they were up to was really different. We had a really good lock on the spectroscopy and then, ultimately, the kinetics. I think Stu was working at lower pressures initially. I don't think he'd tried to jack up the pressures yet. I don't know, something was different. We compared notes on some of the things and went away from that meeting.

About a month or so later, Stu Searles pulled out all the stops on his e-beam gun and got XeBr to lase. It was a curiosity that XeBr was the first of those molecules that lased. It has never really seen much use anywhere because it's terribly inefficient.

Q: XeBr lased in the spring of 1975?

Ewing: Stu Searles did his work around May 1 of '75. He and Hart beat us—beat us with the wrong molecule—but that's okay. So, Stu Searles and Hart did XeBr; we heard that and said, "Now why is it that their e-beam is pumping it, and we can't get xenon fluoride to go?"

And basically the challenge came down from above and from amongst ourselves: "Why is it that NRL's e-beam can do that and our e-beam can't?" We went in and changed our anode/cathode spacing and ripped out half of the foil supports in our hibachi, just threw that all to the wind. We finally had the nerve to turn up all the knobs and work at pressures where before we would have thought we were risking the foil.

It was about May 15th when XeF first went. I could go look at my champagne bottles at home, because I saved them. XeF was the first of ours to lase.

We had previously developed a certain conservatism while we were working with high temperature mercury, because anytime a foil would break, your whole diode would get filled with mercury and it was a mess. You were down for a month almost. But working with room temperature gases, it was much nicer. We put xenon and fluorine in and made those changes, and we got XeF to lase. That was quite a hoot! A week later we did XeCl. Then a week after that we got in our optics to do an old favorite of ours, the nitrogen-oxide gamma bands. We had optics that were coated for making the gamma bands lase. We had this mixture concocted that was argon, nitrogen, and NO, and the NO was supposed to lase. We spent a week working on it, but it didn't lase.

So then we tried krypton fluoride; my model said it was supposed to be at about the same wavelength. We put some krypton and fluorine in and looked at the fluorescence. KrF was by far the brightest thing we'd ever seen. We had been scared of using fluorine at the time. Even though others at AVCO were making chemi-

Figure 13.2 J. J. Ewing (courtesy of J. J. Ewing).

cal lasers with, by comparison, huge amounts of fluorine in them, it took us a while to realize that the amount of fluorine we had was miniscule.

Anyway, we tried KrF, and it fluoresced like mad. Then we put the mirrors on it, and unfortunately the mirrors didn't have real high reflectivity at 249 nm. We had about 20% or 30% output coupling. Even so, the thing just lased like mad. If we had had high reflectivity mirrors on it, it would have just fried them.

Q: How good were ultraviolet optics back in those days?

Ewing: Not very good. They were not very common. That was always a problem. Even windows—we didn't even know how to treat windows. It's important in quartz to make sure you get all the water off, because if you put a little fluorine in there, it etches. We were seeing the effects of photochemical etching.

So we got KrF to go; that was June 5 of '75, I believe. We had several new lasers all in a row there, and that was about the time of the 1975 CLEO meeting. I remember Stu Searles gave his post-deadline talk on XeBr. Right after that I gave a talk on what we had done: KrF, XeF, and XeCl. It was exciting, and I think that accomplishment excited a lot of people. It wasn't just one off-the-wall laser, it was a whole family with promise. We had a good first cut on the kinetics. Charlie Brau realized that electron attachment followed by ion-ion recombination was fast and a dominant pathway.

That is when others started really getting on the bandwagon. Shortly after our work was presented at CLEO, Bhaumik, Ault, and Bradford at Northrop lased XeF using nitrogen trifluoride rather than F_2 as a donor.

The curious thing is that our first excimer-laser publication was about KrF and XeCl, which were really the third and fourth rare-gas halides to lase.

There was another one that lased at that time. There are two kinds of bands in rare-gas halide excimers. There are the ones that go to the main level, the really bright ones that are very sharp because the lower level is kind of flat. Then there are other bands that go to a different final state, bands which are very broad. We always had the idea we'd make the broad ones go.

I had done some calculations and some ruminations on that. Now XeI had the brightest broad bands of all of them in its emission spectrum. So we placed our bets on XeI. We concocted a mix of argon, xenon and I_2 or hydrogen iodide, I forget. We put it in, put on the same optics that we had used for XeF, and zapped it. Jim Dodge, our technician, got it to lase on the first shot! But the light came out at a wavelength that didn't really look like where it should be. So we all celebrated and had our by-now-traditional bottle of champagne. Then we asked, "What is it that's really lasing here?"

The lasing spectra didn't have the broad structure we expected from XeBr. It had all these fluctuations in it. "What is it?" we asked. "Well, let's take out the xenon." We took out the xenon . . .

Q: And it still worked?

Ewing: It still worked. Swell . . . but without the xenon in it, it wasn't XeI lasing. It shouldn't be argon iodide either, because I had done some calculations that said ArI wouldn't have a bound excited state. So the only thing it could plausibly be was something related to I_2. We deduced that it was a state in I_2 that was probably formed by ion/ion recombination of an I^+ and I^- that somehow got made. So we had that I_2 laser in there also. A fourth bottle of champagne and a fourth paper. Subsequently Br_2 and F_2 were lased by various workers on similar transitions.

Q: You've mentioned some of the people who were involved in early excimer work. Who else contributed?

Ewing: Before the rare-gas halide work, SRI (Don Lorents, Dave Huestis, Don Eckstrom, Bob Hill, etc.) had done some very nice kinetics on Ar/N_2 and Hg_2 and HgXe. IBM and Los Alamos also worked on rare-gas dimers. Interest in rare-gas halides blossomed in the summer of 1975. At that time Northrop (Earl Ault, et al.) and Charlie Brau and I at AVCO were pushing as hard as possible to understand and increase the intrinsic laser efficiency. Intrinsic KrF efficiencies of over 10% were reported.

Joe Mangano and Jonah Jacob were the first to demonstrate the discharge pumping of KrF in an e-beam-sustained discharge. Joe and Jonah worked hard on understanding these discharges. Ralph Burnham and Nick Djeu were at NRL along with Searles and Hart. Burnham apparently was the one who said, "Why don't we modify an old Tachisto CO_2 laser, give it a faster circuit, and see if we can make it into an excimer laser?" I think he and Nick got XeF to go first. That was a very, very significant thing. It got excimer lasers to the point where they were in principle no longer a laboratory curiosity, where they could be avalanche-discharge pumped. Burnham and Djeu did that in the summer of 1975. Dick Airey was their supervisor at the time. Much of this work was first reported at Woods Hole in September of 1975. That meeting was a watershed in spurring further work.

Q: What about the Sandia people; what was their contribution?

Ewing: They got involved in the summer of '75 after CLEO. They first talked about their effort at the Woods Hole meeting. They were the first to scale a number of these devices to a large number of joules. In fact, the first time argon fluoride was made to lase, sometime during the winter of '75–'76, they whacked out over 100 J from it. It was the first time the ArF spectrum was even recorded. To me that always was a mind boggler. But they had this huge e-beam—3,000 J at a megavolt or something like that. They really put it to those molecules. That's what Kay Hayes and Gary Tisone were working on. But those people at Sandia didn't just come out of the woodwork, you know. They'd been e-beam pumping lasers for a long time. Wayne Johnson and Jim Gerardo were the big honchos in that effort. Those guys had done a very nice job on xenon excimers. But there were a lot of people involved, and I can't possibly recite them all. In the xenon vacuum ultraviolet excimer work, Charlie Rhodes, Paul Hoff, and Buddy Swingle at Livermore come to mind. John Murray and Howard Powell were also at Livermore, working on rare-gas oxides.

Q: To what extent were the Soviets involved in the development of excimer lasers?

Ewing: I think Danilychev had worked on e-beam pumped xenon, but they were really behind the US work on rare-gas halides. Even a year and a half after we had come out with the first rare-gas halide reports, I know Bill Krupke visited Russia and found there wasn't anyone there doing anything on it. They seemed to be about a year or so behind. In fact, the beauty of the Russian work was that they were behind and didn't read a lot of our papers, so they didn't have some of the prejudices that we had. One of our prejudices that came from our early work was that XeCl was not efficient. We thought XeCl was not so interesting because we were using Cl_2, we were using it very rich in halogen donor. The Russians didn't know about this and simply ignored it. They went out and made XeCl work better in a discharge laser than anyone in the United States had been able to. So the word came back to us, "Maybe we ought to look at XeCl again."

Q: Were there any Europeans involved in the early work?

Ewing: The first publications on gaseous rare-gas halide emission was by Golde and Thrush in England, but they were doing some physical chemistry experiments, and the possibility of lasing wasn't considered. They saw argon chloride emission at 172 nm, and they did their work just slightly before Setser. So if you look at who is the first person to mix some electrically excited xenon or argon or whatever with a halogen donor, it was really Golde and Thrush. And at AVCO, our a initial question on that was, "Why is the spectrum where it is and not so broad?" We went on to making lasers from there.

Q: Have you been impressed by the development of commercial excimer lasers?

Ewing: Yes. There clearly is a good market and the technology continues to improve.

Q: What has most surprised or impressed you about laser development so far?

Ewing: I have been most surprised at how the United States has squandered technological leads and innovations across the board in many segments of laser technology. The focus of so much of the development has been shifting and sporadic. As a result, we have not been able to make the long-term transition from novel research and development to mature technologies, which benefit to the economy in general, and take advantage of the enthusiasm and sound results of the initial breakthroughs described in this book.

We need to take a longer-term view to exploit basic advances properly. Unfortunately, this has not been the case.

Q: What new directions do you expect laser technology to take in coming years?

Ewing: The growth of all solid-state technology is an irreversible trend. I believe that as diode arrays become lower priced, there will be a gradual decrease in the use of other sources. This will be especially true for visible and mid-infrared lasers in the 10 to 100 watt range. I believe CO_2 lasers will always have a niche, as will high-power excimer lasers. Moderate power excimer lasers and dye

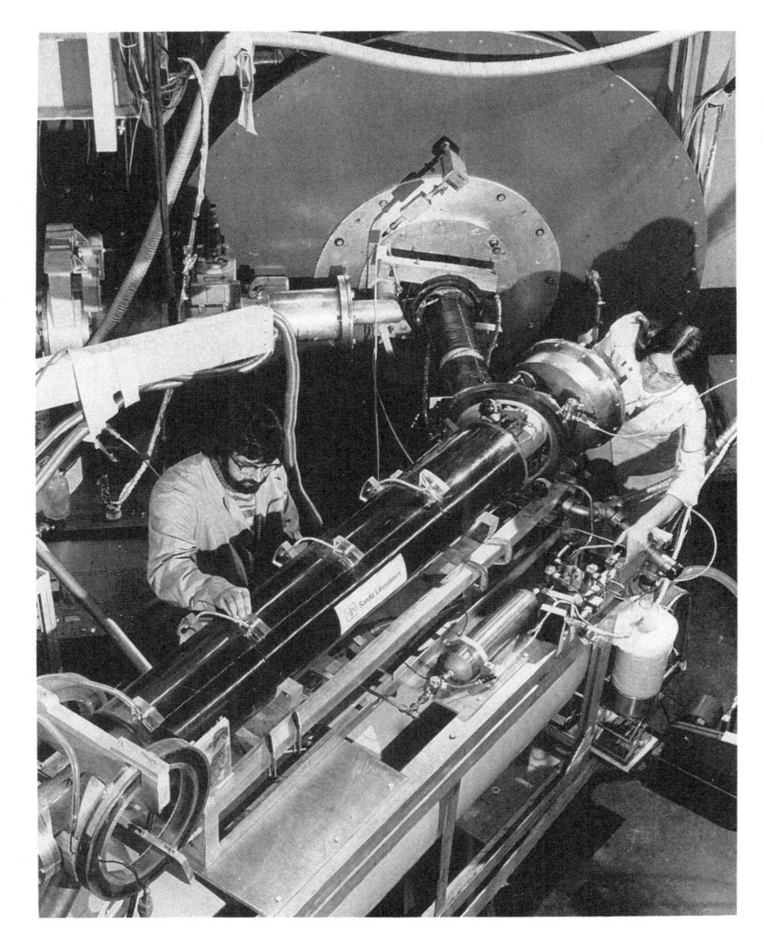

Figure 13.3 Gary C. Tisone and A. Kay Hays of Sandia National Laboratories in Albuquerque used the 7.7-kilojoule electron-beam generator in background to obtain the first laser action from argon fluoride at 193.3 nanometers—an impressive 92-joule pulse. They also increased output energy of krypton fluoride 20-fold (courtesy of Sandia National Laboratories).

lasers will be displaced in time by all solid-state devices, which is not something I fully believed four or five years ago.

Q: What are you working on now?

Ewing: My position at STI now is mostly managerial rather than technical. The technical activities that most excite me, and to which I can contribute, are in the area of tunable solid-state lasers. We have been working in this area for several years, developing titanium-sapphire lasers. They are very nice tunable sources in the near infrared, and in the visible and ultraviolet with harmonic generation. We are well along on some exciting developments in optical parametric oscillators, with the goal being solid-state sources tunable over a wide range of the spectrum. We are just embarking on some exciting work on high-power diode-pumped neodymium lasers.

We also are still doing state-of-the-art development of excimer lasers. We recently built a 2000-hertz KrF laser that can push up toward 100 watts, which is a very interesting machine.

An earlier version was published in *Lasers and Optronics* ®
(formerly Lasers and Applications) a Gordon Publications, Inc. publication.

JOHN M. J. MADEY

The Free-Electron Laser

John M. J. Madey received his bachelor's and master's degrees from the California Institute of Technology in 1964 and 1965, and a doctorate in physics from Stanford University in 1970. He stayed at Stanford after graduation, laying the groundwork for the first demonstration of a free-electron laser amplifier on January 7, 1975. That experiment amplified the beam from an external carbon-dioxide laser. His group later demonstrated the first free-election laser oscillator, which operated at 3.4 micrometers. He also participated in the first demonstration of a visible free-electron laser, conducted at the University of Paris-South in Orsay using the ACO electron storage ring.

He continued free-electron laser research at Stanford, becoming a research professor of electrical engineering and high-energy physics before he left in 1988 to join the physics faculty at Duke University. His current interests include short-wavelength free-electron lasers and the development of laboratory scale free-electron laser technology for applications in medicine and basic research. He is director of the Free-Electron Laser Laboratory at

Figure 14.1 John Madey (left) and Luis Elias (right) work on an early free-electron laser experiment at Stanford University (courtesy of Stanford University).

Duke, which will house a dedicated one-billion-electronvolt storage ring for short-wavelength FEL research. To facilitate interdisciplinary research, the Free-Electron Laser Laboratory is physically located between the physics department and the research laboratories of the Duke Medical Center.

C. Breck Hitz conducted this interview on April 1, 1985, in Madey's office at Stanford University. It was updated in May 1991.

Q: When you were in graduate school, quantum electronics was a brand new field. What led you to get involved with it?

Madey: Well, I took a bachelor's in physics, but even before then I can remember really being fascinated by electron tubes and microwave devices. I guess I'd been into ham radio fairly deeply in high school. As an undergraduate at Caltech, I spent a couple of summers working in a high-energy physics group and getting some exposure both to particle physics and accelerator physics. That was interesting, but by the time I had been through a couple of summers on that and was ready to pick a graduate school, I'd kind of decided that high-energy physics was on too large a scale for me. I decided to look for something that you would do in a normal size laboratory. This was 1964, and lasers were coming on pretty strong then, so I decided to study quantum electronics.

Q: You were doing graduate work at Caltech?

Madey: Right. I ended up working for Amnon Yariv for a year there, and getting a master's in quantum electronics. Then, for various reasons, I figured I had missed the wave in quantum electronics and that things were starting to get a little bit dull. Looking back, that was probably a mistake, but I had done some reading on thin films and low-temperature physics, and I decided that maybe that was really the place to make a career. So I signed up with Bill Fairbank at Stanford and worked for the next five years in low-temperature physics in his group. That's where I got my doctorate.

Q: Did the idea of a free-electron laser come to you somehow while you were doing low-temperature physics?

Madey: Actually, the idea came to me before that. I was taking Yariv's quantum electronics course at Caltech in 1965, and the thought occurred to me that the physics of the atomic or molecular laser was not really unique to an atomic or molecular transition. It seemed to me that if one had a free-electron transition of some kind, it could also—in principle, anyway—be capable of yielding gain. The example that came to mind in 1965 was Coulomb scattering from a nucleus, either in a plasma or a solid, where you could generate a photon as a result of the collision.

The problem appealed to me because it involved a new laser theory. It was a new description of how one might be able to generate radiation, be it microwave or light radiation, from an electron beam. And electron beams, of course, were a long-standing interest of mine.

Q: What was Yariv's reaction to the idea?

Madey: Well, Amnon may not remember it, but I remember it very clearly. We spoke after this particular lecture, and I said, "Look, suppose you had an electron beam and you could scatter the electrons from a lattice, from the nuclei in a lattice, and you'd get some radiation in the final states. Isn't it true that you might be able to get some gain out of that?" And he said, "Oh, yes, that's right, that would work." And, in fact, he gave me a reference to one of his former colleagues at Bell Labs, Dietrich Marcuse, who had looked at that phenomena in maybe 1962 or '63, and used it as an example in his book. I looked it up and sure enough, there was gain, although not a lot of it. There were enough nonlinearities present to give it a fairly low saturation threshold. Besides that, it wasn't really clear that you could have a beam of free electrons running through a crystal without scattering from the phonons.

Q: Should credit for the original concept of a free-electron laser go to Dietrich Marcuse?

Madey: My guess is that, if you looked further, you would find others to whom the idea had occurred, and who had made at least some back-of-the-envelope calculations.

Q: So one can't really say where the idea of the free-electron laser originated?

Madey: Well, the first reference that I've seen to the possibility, indeed the inevitability, of amplification in the scattering of free electrons was a 1964 paper by Dreicer at Los Alamos. He was looking at the attainment of equilibrium in plasmas. He pointed out that if you just looked at spontaneous scattering events—in particular, at thermalization of electrons in an intense radiative environment—the cross-section for normal Thompson scattering would be too low. He realized you had to include stimulated radiation in the problem to get realistic thermalization times. And if you had stimulated radiation, you could get gain.

So, that was the first published reference I've seen to amplification from free-electron transitions. As I said, that was in the context of a plasma-physics problem.

Q: You were saying that you left quantum electronics and came to Stanford to do low-temperature physics with William Fairbank. How did you wind up in laser physics again?

Madey: The discovery of the Gunn effect, in about 1966, perked my interest in free-electron lasers. You take a piece of bulk gallium arsenide, bias the sides relative to each other, use non-rectifying contacts, and you find that the thing oscillates. In fact, we get quite a lot of power out of those things; people don't make klystrons anymore. But it seemed to me, my God, this is really it! Here's your solid-state free-electron laser. This is really hot! If you recall, at first no one really knew what was actually going on in the Gunn effect. Nobody understood the microscopic physics, or even what was causing the oscillation. Now we know it's the formation of the low resistance domain as electrons scatter into a Gunn minimum, which then propagates across with some characteristic transition time and reforms. That's what's responsible for the oscillation. But at that time it seemed to me, gee, this might really be the microwave free-electron laser.

Well, the numbers didn't come out right when you tried to explain the Gunn effect that way. And then, within a relatively short time, someone provided the correct explanation. But the excitement had restored my enthusiasm to the point where I figured I'd keep kicking around the ideas. Bill Fairbank was very understanding about this. Even though I was working on a low-temperature physics problem, I kept stirring the thoughts on free-free interactions.

Q: How long did this thought-stirring process last?

Madey: After one or two years, I came to understand more thoroughly why Coulomb scattering wouldn't work. Basically, the reason the gain was so low was that the transition probabilities for emission and absorption were almost identical. So there was very little net gain available. That's because the emission spectrum for Coulomb scattering is so broad. If you could get narrow-band emission from some sort of free-free scattering, your chances of gain would be a lot better. Once I realized that, I immediately thought of Compton scattering.

There was another important thing I'd found out about several years earlier. The Weizsacker-Williams approximation states that electrons moving at relativistic velocities cannot tell the difference between a traveling electromagnetic wave and a static electric or magnetic field. That is, even if you start with purely a magnetic field, the Lorentz transformation always generates an accompanying electric field in the electron rest frame.

Things really clicked at that point because now we could increase the amplitude of the scattering—use a superconducting magnet or a huge permanent magnet to create a much stronger equivalent field than you could get in a real electromagnetic wave. To me, it looked like we could probably get a reasonable amount of gain, even in the visible.

Q: While you were figuring all this out, were you also working on your thesis in low-temperature physics?

Madey: Right. My thesis actually included both the low-temperature physics problem and the FEL [free-electron laser] theory. By 1969 I had a valid gain equation for an FEL and criteria for e-beam current, quality, and so forth. And I had some idea what we'd have to do to make an experiment work. In 1971, we started thinking seriously about building a free-electron laser. We had to generate strong, static, and periodic magnetic fields. And we had to have a good source of electrons at the proper energy, 20 to 50 MeV for emission in the infrared.

Q: How long did it take to get the first FEL to work?

Madey: We started research on the program in the fall of 1972, and that ran for four years before the first oscillator was demonstrated. But we demonstrated amplification, using a CO_2 laser as the driver, in 1975.

Q: Did you have difficulty getting the work funded?

Madey: Well, those were the late '60s and early '70s, the golden years of physics research. We were running on about $200,000 a year from the Air Force Office of Scientific Research. That was quite an adequate budget in those days. In fact, we weren't sure we could spend it all when we started.

Q: Did you encounter skepticism about the FEL from your colleagues?

Madey: Not at Stanford. We were fortunate that a new superconducting accelerator had just come on line in the laboratory, and it was exactly what we needed. It had the right energy range, very low energy spread, excellent beam quality, and just enough current to bring the system above threshold. In fact, the laboratory staff and administration were fascinated about having an experiment which seemed uniquely suited to the capabilities of this new accelerator. So, we got good cooperation from the laboratory in setting this system up and in securing running time for it.

Q: I recall a lot of people were running around saying that the FEL wouldn't work.

Madey: Ah . . . well, golly, I won't mention names, but I had people visit me who insisted that you would never see the narrow emission spectrum needed to make the laser work. Even after we had measured it, they insisted that it couldn't be correct.

But the important thing is that we had the commitment of the Air Force Office of Scientific Research, and they were willing to play out the string to see what we could do. We thought we understood the physics pretty well, you know.

Q: Even so—were you surprised when the FEL first worked?

Madey: Nope. Things were very close to what we expected.

Q: Were FELs being investigated anywhere else? Was there any concern that another group might operate an FEL first?

Madey: I think that, outside Stanford, there was, if not a wildly negative view, a sufficiently pessimistic evaluation of the possibilities, that we really didn't have to worry about competition.

Figure 14.2 John M. J. Madey today (courtesy of John Madey).

Q: Who contributed to the development of the FEL?

Madey: Three people in particular really deserve mention. One is Dick Pantell, who had done a calculation on the Compton scattering problem for free electrons in a real microwave field. He published his results with Soncini and Putoff in one of the IEEE journals in 1968. He had the right numbers for scattering from a microwave field.

The second fellow is Hans Motz, who was a research fellow at Stanford in the mid-1950s. He later moved to Oxford University in England, and died in August 1987. Of course, he wasn't motivated by laser theory, because that hadn't been invented then. But he noted that if you ran an electron beam through a periodic magnetic field, you would get a nice, narrow emission spectrum. He saw that if you took advantage of the bunching that was inherent in the electron beam from a linear accelerator, or if bunching could be induced in a traveling wave tube, you could get very high power output. Even if you got only spontaneous radiation, that looked interesting as a source. But, if you could get the electrons to radiate coherently in the undulator, it could really be a very intense source. So he built an undulator. He ran an e-beam through it, in maybe 1956, and saw this rainbow of colors. It was a really nice experiment.

The third individual is a chap by the name of R. M. Philips, who was working in GE's Palo Alto Research Laboratories in the late 1950s and early 1960s. He had gotten interested in the periodic-magnet geometry also. He made numerical calculations of the bunching caused by an input electromagnetic wave. He built an amplifier that ran at microwave frequencies and generated tens of kilowatts of average power at wavelengths around a centimeter or two.

Q: Did Philips and Motz know about each other?

Madey: They may not have. I believe Philips was supported by a classified contract, so the full details of his research may not have been available in the open literature. I haven't seen cross references in their publications, and it may well be that they were working completely independently.

Q: Were other people involved in the early work with FELs?

Madey: There was another analysis, one that's only recently become known, which I think is really remarkably prescient and complete. It was prepared by Ken Robinson, who was an accelerator physicist at the Cambridge Electron Accelerator in Cambridge in the late '50s. He had developed a good classical model for the interaction, and he got the right numbers for gain. He prepared a rather detailed description of an infrared or visible oscillator and amplifier. The tragedy is that the Cambridge accelerator was closed down in 1966 or 1967. That accelerator was Robinson's baby, I guess. He was never professionally employed again. The analysis of the free-electron laser was discovered in a trunk of his possessions after his death.

Q: It's a sad story.

Madey: It really is. The other tragedy is that, for its time, the Cambridge Electron Accelerator was a remarkably advanced machine. It might still be performing useful work today, certainly as a synchrotron radiation source, and perhaps even for FEL research. But apparently there was an agreement that called for the Department of Energy to return the site to its original state if funding was ever terminated. So the whole thing was knocked apart. Not a piece remains.

Q: Well, that's really a discouraging story! What about the people who worked with you at Stanford. Who were they and where are they now?

Madey: We had a remarkable group of people working on that experiment. In particular, there's Luis Elias, who now heads up the FEL program at the University of Central Florida in Orlando; Dave Deacon, who has his own company in Palo Alto; Todd Smith, who's continuing to push ahead in linear accelerator technology at Stanford; and Bill Colson, who developed the classical theory for the FEL at Stanford, and is now a faculty member at the Naval Postgraduate School in Monterey.

Q: Where do you see free-electron lasers going in the next decade or so?

Madey: It's a hard question to answer, particularly for large systems. FELs have benefited from a lot of enthusiasm over their potential, as well as from perhaps lack of full knowledge or understanding about the limitations and the technical difficulties. So, I have some concern about unanticipated problems that could slow down development of the big FELs.

As far as small FELs for scientific research are concerned, we've got a long-enough shopping list to keep the technology active and expanding for quite some time. One of my active programs is the development of a small-scale infrared source for production of modest power levels, ten to a hundred watts. Further in the future, we hope to put together an extreme ultraviolet source operable in the region below a thousand angstroms, perhaps to wavelengths as short as 200 angstroms.

Q: What sort of power would you expect in the extreme ultraviolet?

Madey: It would be mirror-limited. If the only problem were the e-beam and the magnet, you might talk about powers of perhaps 100 watts to a kilowatt, from 1,000 angstroms to 200 angstroms. But I think mirror limitations will keep it far below that.

Q: What are you doing now at Duke?

Madey: The university has completed a dedicated FEL laboratory with 50,500 square feet of space. It will house a one-billion-volt storage ring, as well as research facilities to develop and utilize smaller long-wavelength free-electron lasers and free-electron laser instruments for basic and medical research. The building cost $5 million, and the government investment in the free-electron laser hardware inside to date totals about $25 million.

We already have used our infrared FEL to study the wavelength dependence of tissue removal in surgery, evaluate the removal of gallstones with picosecond pulsed lasers, and study energy transport in proteins at the subcellular level. When our storage-ring laser is completed, we plan to generate ultraviolet and x-ray laser beams for microprobe and holographic imaging, and the analysis of cellular function and pathology. Although these concepts have been demonstrated using existing light sources, they have not been reduced to practice because suitable coherent sources have not been available at short wavelengths.

Q: Would you do anything differently if you had it all to do over again?

Madey: I sure would. We spent a long time getting deeply into the laser physics from both the theoretical side and the experimental side. We kept the first experiment running for almost five years after the first demonstration because there were enough things that we didn't understand or wanted to get better diagnostics on. I think that was really a mistake. I think we made a mistake in not trying to develop less expensive ways of making FELs. We had an advanced superconducting magnet and a superconducting accelerator, all of which were fine in terms of getting a clean beam and getting good data.

But, boy, we sure would have been better off if we'd stepped back and said, "Well, look, what could we do that would cost us, say, $100 a day to run instead of $10,000, even if it doesn't have some of these nice characteristics?" Because we knew enough at that time that we could have pushed towards such a system and had a small scale FEL operating before the end of the '80s. And FELs would be better off now if we had done that. In every successful area of laser technology, somebody's come along and figured out a cheap and dirty way to make the laser.

Q: Like developing discharge-pumped excimers?

Madey: Exactly. And we put that off, I'm afraid, for five years longer than we should have.

An earlier version was published in *Lasers and Optronics* ®
(formerly Lasers and Applications) a Gordon Publications, Inc. publication.

DENNIS L. MATTHEWS

The X-Ray Laser

Dennis L. Matthews majored in physics as an undergraduate at the University of Texas at Austin, and received a doctorate in atomic physics from the same school in 1974. After graduation, he joined the E Division at the Lawrence Livermore National Laboratory, where he continued his thesis research on highly ionized atoms. In 1979 he joined Livermore's Inertial Confinement Fusion program, where he studied plasma conditions with x-ray spectroscopy. He considers himself an experimentalist, with fields of expertise including collision phenomena of heavy ions and atoms, Auger electron and x-ray spectroscopy of ions and atoms, and spectroscopy of highly ionized plasmas. He has more than 100 publications in the scientific literature and holds many patents.

Matthews first became involved in x-ray laser research at Livermore's E division and gave his interview seminar for the inertial fusion program on x-ray lasers. As an experimentalist, he later teamed with theorists Peter Hagelstein and Mordy Rosen to demonstrate laboratory x-ray lasers, powered by the glass fusion lasers at Livermore. (His group overlapped with the Livermore team

Figure 15.1 Dennis Matthews in his office at Livermore (courtesy D. Matthews).

which developed x-ray lasers driven by nuclear explosives, research that remains under tight security restrictions as part of the nuclear weapon program and thus can be covered only in passing in this interview.)

In 1982 and 1983, Matthews assembled a team of researchers and was the project leader for the experiments that eventually demonstrated the selenium x-ray laser on Friday, July 13, 1984. While other groups earlier had claimed to have observed x-ray gain, the Livermore demonstration was the first to be widely accepted as solid evidence for x-ray amplification.

Matthews is now group leader for development of laboratory x-ray lasers and their applications in Livermore's inertial confinement fusion (ICF) program. His group has continued pushing the frontier of x-ray laser research. They now hold the record for shortest laboratory x-ray laser wavelength, 35.6 angstroms, significantly shorter than the 206 and 209 angstrom selenium wavelengths in the original demonstration. Laser development is coordinated with the Biomed program designed to use x-ray lasers to record holograms of living cells. The group's next goal is demonstration of "tabletop" x-ray lasers, with drivers much more compact than the massive Nova fusion laser used in most of their experiments.

In 1985 Matthews and his research team shared the Department of Energy Award for Research Excellence for their 1984 experiments. He has been elected a fellow of the American Physical Society for his pioneering work on x-ray lasers.

This interview was conducted by Jeff Hecht at the Marriott Keybridge Hotel in Vienna, Virginia, on February 22, 1990.

Q: How did you get interested in science?

Matthews: When I was six or seven, I read comic books, which contained a lot of scientific information. That grew into, "Dad, can I have a chemistry set? Dad, can I have an electrical set?" I got more serious in high school. Before that, you don't know enough to ask the right questions, but then life begins to sparkle.

I was interested in astronomy, but that wasn't an undergraduate major at the University of Texas at Austin. I almost said I'd rather be a dentist, but decided to take physics. For a while, I had an ego trip about knowing things that other people didn't know, which turned into genuine curiosity about things that no one knows.

I like to see things happen, to interpret them and to thumb my nose at theorists. I don't know why, because I did well in theory courses. Maybe I like the empirical method. One of my professors told me—I didn't believe it then and I still question it—that I had the golden touch, and when I work on things, they work. Probably I either was cautious enough to pick something that was going to work anyway, or could get the right people to work with me. You have to be careful not to let it go to your head, but sometimes you think, "I'm supposed to do this."

Q: X-ray laser research started in the 1960s. How did you get involved?

Matthews: My doctoral thesis was on Auger electrons, which are ejected by very highly ionized atoms when one electron picks up energy released by another electron as it drops to a vacant inner energy level. I got interested in x rays as the alternative to Auger emission. In the mid-1970s I went to work for Dick Fortner at Livermore, who justified studying highly ionized atoms by talking about making a population inversion. We joked that making an x-ray laser would cost an arm and a leg. Then I got interested in ions in high energy–density plasmas, and became the x-ray spectroscopist for the laser fusion program.

Working with Fortner, I met the Star Warriors [the subjects of William Broad's book, *Star Warriors*], Lowell Wood, George Chapline, Tim Axelrod, and others. Peter Hagelstein, who was a student, and Tom Weaver came about the same time. They were all theorists, and had thought quite a bit about using an optical laser to make a plasma dense enough and far enough from equilibrium to be an x-ray laser. In 1972 Lowell and George had published an x-ray laser proposal, and later they petitioned for experimental time. The Russians—A. V. Vinogradov, A. Zherikhin, and others—also published x-ray laser proposals. At any rate these Livermore theorists got wind of my talk about making x-ray lasers, and realized they had a friend and an experimentalist in the right position to perform experiments in the Laser Program.

Q: In 1972 the University of Utah had claimed to have made an x-ray laser by zapping gelatin with a pulsed laser, but no one could duplicate it. What happened?

Matthews: They had a dark spot on film which might have been caused by a beam. That was a necessary but not sufficient condition for demonstrating an x-ray laser, and it proved the downfall of other people, too.

Q: What were your original plans?

Matthews: We wanted to pump an x-ray laser with [the 10-beam] Shiva [glass laser built for inertial confinement fusion research] as part of Peter Hagelstein's thesis. In the late 1960s, Michael Duguay of Bell Labs had proposed what is called a photo-ionization laser, based on selectively removing electrons from the inner shells of neon to make a population inversion. Peter modeled a 230-angstrom laser, using neon sandwiched between other materials, but we got sidetracked. We wanted to make shorter wavelength lasers, and Peter wanted to try an idea proposed in his thesis, pumping x-ray laser levels with emission lines from plasmas produced by other lasers.

The photoionization cross section is very strong if the plasma emission line overlaps the absorption line exactly. In about 1980, Peter suggested that chromium stripped of all but three electrons—lithium-like chromium—could pump hydrogen-like fluorine [stripped of all but one electron] to make an 81-angstrom laser. We built and tested amplifiers that were chromium foils surrounding a gas column containing fluorine. Emission from the chromium plasma had both to ionize fluorine and to excite its laser state, so the foils could only be 50 micrometers apart. Our understanding of the physics and our modeling said this had to be very, very good for the fluorine to lase before the fluorine plasma collapsed, having been replaced by the imploding chromium foils.

Q: Wasn't the excited-state lifetime very short?

Matthews: The decay rates were horrendous, 10^{12} or 10^{13} decays per second—meaning lifetimes were 0.1 to 1 picosecond. We had taken for granted that we had calculated everything perfectly and that it would work according to prescription. However, we couldn't pump fast enough with Novette, the 2-beam version of the [current 20-beam] Nova laser. Our first round of tests showed no evidence of lasing.

After those tests, I went to KMS Fusion in Ann Arbor with Phil Burkhalter [of the Naval Research Laboratory] to measure resonances to very high accuracy. We found an emission line of beryllium-like manganese only a milliangstrom from the 12.643 angstrom transition we wanted to pump in hydrogen-like fluorine. That sounded good, but at the national labs you become vulnerable once you start to fail a little bit, so Peter and I started thinking about some Russian schemes we could also test to increase our chances of success.

Peter and I determined that $3p$–$3s$ lasers in neon-like ions looked pretty robust, although their wavelengths were not that short. That inversion depends more on atomic physics than on reaching horrendous pump brightness while having the plasma hold together just right, and it didn't require a separate pump plasma. Neon-like krypton had been modeled, but it hadn't worked on a big pulsed power machine at Physics International in San Leandro [California]. I didn't want to mess with gas targets, so I turned to selenium, the usable solid with the nearest atomic number. Preliminary tests at KMS even showed we could get the right ionization balance.

Meanwhile, Mike Campbell and Mordecai Rosen of Livermore suggested exploding foil amplifiers, so we could make plasmas with a very gentle density gradient perpendicular to their length. The idea of hitting a sheet had a simple elegance to it. The plasma would be much bigger than our original 25- or 50-μm gas targets, which didn't stay together very long. We started with 750 angstroms of selenium on a plastic backing, so the laser pulse would eat through the foil at its peak power. We didn't have enough energy at KMS to make an amplifier, but we did make neon-like ions.

I went back to Livermore for our second round of 10 Novette shots. I thought neon-like selenium was the scheme most likely to succeed, and I started with the theoretically predicted parameters, planning to change them in empirical steps. We did the first few experiments with the prescribed conditions, and the results didn't look good, with no evidence of the laser transitions.

The critical step was doubling the pulse length from 250 picoseconds. One colleague called it "a ridiculous excursion in parameter space," and almost talked me out of doing it, but I decided to go ahead. I thought the plasma looked a little underionized, and wondered if we understood how long it took to burn through the foil. We tried it on Friday, July 13, 1984, and now my colleagues always love to do shots on Friday the thirteenth. The [x-ray laser] lines practically burned a hole in the film [laughs], and the rest is history.

Q: Was that your tenth shot?

Matthews: We had shot a few more than 10 at that point, but certainly not 20. We even went back and saw that it was very, very faint with shorter pulses. The signal grew exponentially with amplifier length. It was much brighter than nearby lines not expected to lase. Later we showed that it was a beam and that we could get a spot. We're well over a thousand shots now.

Q: Was it smooth sailing from that point?

Matthews: Not really. These things shoot beams out both ends, so we put diagnostics on each, but at first we didn't see much. One spectrograph had low spectral resolution, so it was very hard to see lines. The other one had good resolution so we tuned it to the $J = 1 - 0$ transition, which theory predicted would have all the gain, but we had trouble with it. I finally said, "I'm taking it off," and put on an instrument called McPigs, an acronym for MicroChannel Plate Intensified Grating Spectrometer. It showed us strong $J = 2 - 1$ lines at 206 and 209 angstroms, which ironically were outside the range of the other spectrograph. The $J = 1 - 0$ line, which we had expected to be a monster, was very weak, which has never been explained.

Q: Other people also were trying to make x-ray lasers. The work given the most credibility before yours was a 1980 report by Geoff Pert's group at the University of Hull in England. What did they do?

Matthews: They used exploding carbon fibers to produce a population inversion by three-body recombination, first stripping all the electrons from an atom, then recapturing one in a process that involves a second electron. It preferentially fills higher states. The plasma has to be very very hot to get a high ionization rate, but if you get too hot, the recombination rate, the density, and the gain drop. What makes Pert's idea brilliant is that the exploding fiber gives a cylindrical expansion that cools the plasma much faster than the planar expansion of a foil, enhancing recombination probability, and not populating the laser ground state.

At that time Geoff didn't have sophisticated diagnostics, just a time-integrating spectrograph. Spontaneous emission from hydrogen-like carbon ions [C^{+5} with one electron] lasted for 10 nanoseconds and tended to dominate the signal, but lasing at 182 angstroms lasted only 10 picoseconds or perhaps 1 nanosecond, and was thus very hard to prove. In addition he, like Pierre Jaegle in France in the late 1970s, had fairly short plasma columns, so even with high gain there was not a big difference between on-axis and off-axis emission.

Q: Another experiment got extensive play in the general press. In February 1981, *Aviation Week & Space Technology* reported that Livermore used a nuclear explosion to power an x-ray laser at the Department of Energy's Nevada Nuclear Test Site. Nobody could talk about it publicly then, but in 1984 Livermore was careful to call your work the first "laboratory" x-ray laser. Was there much overlap between the laboratory and bomb-driven x-ray lasers? What can you say about that program now?

Matthews: Peter Hagelstein, Lowell Wood, George Chapline, and Tom Weaver definitely overlapped, and some experimentalists also were involved in the Nevada experiments. I can say that there is a nuclear-bomb-pumped x-ray laser, and that research was done on it, but I can't confirm or deny what *Aviation Week* reported in 1981. The only other thing I can say is that they made the shortest wavelength x-ray lasers to date, which shouldn't surprise you.

Q: They had a bit more pump energy.

Matthews: Exactly. The two programs helped each other quite a bit. We could do things at a higher repetition rate, and we learned how to diagnose x-ray lasers and how to interpret the data. If the underground program hadn't existed, I don't think we would have gotten enough resources to do the unclassified x-ray laser. Some of the ideas came from the same people, and at times it was difficult to tell which cart led which horse.

Q: How supportive was Livermore management?

Matthews: Early on one man propelled us in the right direction more than anyone else. His name is Roy Woodruff. You may remember him. Since the outset of our project, the present laboratory director, John Nuckolls, has also been a strong supporter.

Figure 15.2 It takes a lot of people to make an x-ray laser; this is the group involved in the Livermore experiments in standard attire, with nary a necktie in sight (courtesy D. Matthews).

Q: Woodruff was the center of a big controversy at Livermore. He apparently was demoted for saying that Lowell Wood and Edward Teller oversold bomb-driven x-ray lasers. He eventually was recognized as a whistle-blower and moved back into Livermore management, after outsiders leaked his personnel files to the press.

Matthews: Early on, we had trouble getting our act together. One day Woodruff, who then was just under the director of Livermore, called Tom Weaver and myself into a room with several others. He said, "There will be a laboratory x-ray laser experiment, and Dennis will do it, and here's the money." Without that, I don't think it would have happened.

Q: How was your program funded?

Matthews: By both the weapons program and the laser [inertial confinement fusion] program. The laser program gave us thousands of Nova shots, which cost some $30,000 apiece. A few years ago, the bomb-pumped x-ray laser program lost interest, but the laser program kept us going. Now we are getting money again from the underground program. They were the real iceberg; we're only a little tip. We spend about $2.5 million a year. The [bomb-driven] nuclear-directed-energy weapon effort is probably $60 or $70 million a year.

Q: The bomb-driven program started before the Strategic Defense Initiative, which funds it now. Who supported it originally?

Matthews: It first came from Livermore's discretionary budget, but after some successes, it got outside money.

Q: You reported your first laboratory x-ray laser at the American Physical Society's Plasma Physics Meeting in October 1984. I recall you had a joint press conference with a group from Princeton University reporting their own x-ray laser. Were you surprised by their work?

Matthews: I had heard of the Princeton work the previous summer, when Szymon Suckewer reported indications of gain at a conference in Boulder. I was surprised they decided to co-announce at the press conference, because I thought they had already claimed x-ray amplification, although not very loudly.

Suckewer, Charles Skinner, and some of Suckewer's students (who now work with us) were exploring recombination lasers using a different idea than Pert, but on the same 182-angstrom carbon line. They made a plasma, then trapped it in a solenoidal magnetic field long enough for recombination to occur. Their wavelength was shorter too, because we got 206 and 209 angstroms from selenium.

Q: Did you have more convincing evidence of laser amplification?

Matthews: At that time, they could not vary plasma length to show how power scaled, but we could, simply by varying the length of the pump laser beam. Instead, Szymon inferred a value for gain from the asymmetry between on-axis and off-axis emission. His gain-length product was three or four then, but now he's gotten up to six or eight, which we had in 1984. The difference in amplification was even larger, because it is the exponential of the gain-length product [$e^{\text{gain} \times \text{length}}$]. Ours was about 1000.

Q: The x-ray laser is unusual because who gets credit for the "first" one depends on how you define an x-ray laser. Pert may have had the first evidence for amplification, the underground group may have had the first x-ray laser, and you and Suckewer's group both observed "first" laboratory x-ray lasers.

Matthews: The best complement we received was from Mike Key, a Briton who eventually worked with Pert to produce their own demonstration of gain with carbon fiber. He said [Nature *316*, p. 314 (1985)] we had the first incontrovertible evidence of x-ray amplification. We put our hordes of people to work on a solid demonstration that was bulletproof, because there had always been a little cloud of doubt over previous work. We were scrutinized in private by [Charles] Townes, [Arthur] Schawlow, and several others.

Q: What have you done since 1984?

Matthews: Our celebration of producing x-ray amplification was short-lived. People were quick to say that "real" x rays were shorter than our 200 angstroms. There's no official boundary, but most people draw the line at the carbon K edge [the longest wavelength that can strip an inner-shell electron from carbon], 43.6 angstroms.

In 1985 Nat Ceglio and I called together people who wanted bright x-ray sources. The biological imaging community wanted more powerful, more coherent, and particularly shorter wavelength x-ray lasers. They gave us ideas about x-ray laser holography and off we marched. First we extrapolated neon-like systems to shorter wavelengths by using heavier elements. In a few months, we got up to molybdenum, which lased at 133 angstroms, but we were running out of power from Nova, and realized we were not going to reach 45 angstroms.

Our next idea was a $4d$–$4p$ transition from ions stripped down to a nickel-like electron configuration. The kinetics were the same as for neon-like ions, but the transition was from the next shell up, and it could give shorter wavelengths with about the same pump energy. Brian MacGowan, Steve Maxon, Peter Hagelstein, myself, and others made nickel-like europium lase right off the bat at 71 angstroms. That was a big step.

My young colleague, Brian MacGowan, has since gone to heavier nickel-like elements, as we did with neon-like systems. Right now the record is tungsten, which lases at 43.16 angstroms, in the domain where everybody agrees we have x rays. We're just writing it up for *Physical Review Letters.* Earlier we reported a nickel-like tantalum laser at 44.8 angstroms, which is the perfect place for holography. We think we can extrapolate that scheme to nickel-like gold, which should lase at about 35 angstroms. [Experiments a few weeks after the interview showed strong evidence for amplification at 35.6 angstroms.] Then we're going to run out of gas with that scheme, but there are plenty of new ideas.

Q: What are others doing?

Matthews: Suckewer is still improving the carbon laser, and he has done some work with heavier elements. Lately he's been studying some "lithium-like" lasers in the 130 to 150 angstrom range. He gets a lot of energy because they emit for a long time, but I don't think he can get much power. He's also looking at some neat new ideas, such as pumping a plasma in his solenoidal magnetic field with picosecond lasers to get down to 10 or 15 angstroms. He hasn't demonstrated anything yet but has been busy building the pump lasers.

We may try pump lasers in the 100-femtosecond regime, to ionize the inner K shell right off a neutral atom and set up an inversion. This is an idea currently being tested by Roger Falcone and his coworkers at Berkeley. It's a self-terminating laser, but who cares? If you can ionize enough atoms, you'll have an inversion long enough to get an amplifier.

Q: Peter Hagelstein is looking at small x-ray laser at the Massachusetts Institute of Technology, as is Suckewer at Princeton. Is anyone else?

Matthews: Some encouraging data has come out recently. T. Hara, a professor at the Institute of Physical and Chemical

Research (RIKEN) in Japan, claims he's pumped a tabletop x-ray laser with a 6-joule neodymium-glass laser, but nobody's duplicated it. Suckewer has reported pumping 182 angstrom carbon lasers with 25 joules, and maybe even as little as 5 joules. My only caution is to remember a warning from Bill Silfvast: "A few gain lengths does not a laser make." I think tens of joules can drive an amplifier, but I want to make sure its output is usable. For the near future, we're looking for simple steps down from Nova's tens of kilojoules to 0.1 to 1 kJ.

We are one of the few fields that wants smarter and smaller machines rather than bigger, and that's a good sign because we're learning how to do make x-ray lasers more efficiently. Geoff Pert and Pierre Jaegle have been learning how to go up from very small systems, while we learn how to work downward from a large pump source. Merging our ideas, I hope, will give us something usable and efficient.

Q: Do you have any particular objectives for x-ray holography of living cells?

Matthews: Our clear objective was to find an important application for an x-ray laser. My colleague Jim Trebes at Livermore and biologist Joe Gray, also at Livermore, want the ability to image living cellular material with 300 angstrom resolution so they can study what it does in normal healthy cells, in unhealthy cells, and in special cases such as when cells are being fed drugs.

Nobody else can look at the living cell. Of course, taking an x-ray hologram destroys it, but until then it's functioning. Electron microscopy has much better resolution, but it requires killing the cell, staining it, crystallizing it, and other things, which leave biologists unsure what's happened to the structure. More important, they can't see how the cell reacts to a stimulus. In principle, a 45-angstrom x-ray laser should give at least 45-angstrom resolution, and perhaps half that, but we think making the hologram and reconstructing the image may degrade resolution to the 300-angstrom range.

The nickel-like laser is almost at the point needed to make holograms. Five years ago I didn't dream we would be this close, but with enough resources, we can do it in the next couple of years.

We've got enough energy, but it has to be fully spatially coherent, and that's a little harder, but not impossible, especially now that we have multilayer x-ray mirrors.

Q: Are x-ray mirrors changing what you're doing?

Matthews: They make life easier, although their reflectivity is at best about 60–70%. A mirror on one end allows two passes and thus higher extraction efficiency from our amplifiers. We also plan to test mirrors on both ends of the amplifier, pumping each time the x rays pass through the cavity. The gain is high enough to saturate a selenium amplifier at 200 angstroms after five or six passes. We're planning tests with the Janus laser at Livermore. It's a dinosaur in our closet, which only puts out 100 joules, compared with 100 kJ from Nova, but it's a lot smaller and cheaper way to work toward "tabletop" x-ray lasers.

Q: What about other applications?

Matthews: We've looked into x-ray interferometry of laser-produced plasmas and studies of transient ionization and excitation phenomena. X-ray lasers should make a whole different kind of plasma than we've ever seen. But it's tough to figure all that out. I asked Townes, "What did you do when people asked you what to do with the laser?" and he said, "I was never so presumptuous as to think I knew. All I knew was that I was going to make a damn good source." Unfortunately, that doesn't get me funded these days, but I like the spirit of it.

Q: What happened to the Russians? They did some early theoretical work.

Matthews: I've never seen any experimental work published, but A. Zherikhin, K. Khoshelev, Vladilan S. Letokhov, A. Vinogradov, and numerous others have all done theoretical research. They are very interested in what we're doing. One can only speculate about their work. I have heard from separate sources that they had either concluded x-ray lasers were impossible, or that they could no longer discuss them since their work may have become secret in the late 1970s. We've been a little frustrated, though, because they have nothing to offer at our conferences. However, I don't feel too badly because I think they had most of the original theoretical ideas, but they simply lacked the technology to test them.

Q: How many people have been involved in your work?

Matthews: We've always numbered at least 10, although only a few people were the driving forces. John Emmett once warned me: "Don't ever appear on anybody's management chart, or they'll snip you off and you'll disappear." For a long time we were just a revolutionary group—a "skunk works"—interested in x-ray lasers, and no one could stamp us out, because we didn't exist formally. Unfortunately, that's going away, and we have to devote more effort to what everybody else has to do, getting money and justifying our existence. Since 1984 I've had less time to do research, but the notoriety of our work has attracted some very good people, and they're probably doing an even better job.

Q: Looking back, would you do anything differently?

Matthews: No. I was fortunate to be in the right circumstances at the right time to do the x-ray laser experiments. I do feel badly about how much energy we have to spend fighting for money. I'm not experienced enough to know if that's normal or if we had to jump through extra hoops because of the x-ray laser's connection with political and strategic issues. Sometimes it seemed that it didn't matter how good a job we did; we were going to be done away with just because we weren't in fashion anymore.

For Further Reading

Other publications and recollections of laser history

L. Allen, *Essentials of Lasers* (Pergamon Press, Oxford, 1969) [reprints historic papers].

George C. Baldwin, Johndale C. Solem, and Vitalii I. Gol'danskii, "Approaches to the development of gamma-ray lasers," *Reviews of Modern Physics 53* 687–744 (Oct. 1981).

Robert Bellinger, "Gordon Gould's sweet victory," *Electronic Engineering Times* 64–68 (Apr. 18, 1988).

Mario Bertolotti, *Masers and Lasers: An Historical Approach* (Adam Hilger Ltd., Bristol, UK, 1983).

Bigelow, Holly, and Susan Lamping, "Women in lasers," *Lasers & Applications 5,* 1, 59–65 (January 1986).

Nicolaas Bloembergen, "Nonlinear optics and spectroscopy," (Nobel lecture), *Reviews of Modern Physics 54,* 3, 685–693 (1982).

Nicolaas Bloembergen, "Nonlinear optics, past, present, and future," *IEEE Journal of Quantum Electronics QE-20* 556–558 (1984).

Nicolaas Bloembergen and Arthur L. Schawlow, "Bloembergen, Schawlow reminisce on early days of laser development," *Optics News* 13–16 (Mar./Apr. 1983).

William Broad, *Star Warriors* (Simon & Schuster, New York, 1985).

Joan Lisa Bromberg, "Research efforts that led to laser development," *Laser Focus/Electro-Optics* 58–60 (Oct. 1984).

Joan Lisa Bromberg, "The construction of the laser," *Laser Topics 7* 2–6 (Nov. 1985).

Joan Lisa Bromberg, "Engineering knowledge in the laser field," *Technology and Culture 27* 798–818 (1986).

Joan Lisa Bromberg, "The birth of the laser: developments in the United States," *Physics Today 41* 26–33 (Oct. 1988).

Joan Lisa Bromberg, *The Laser in America 1950–1970* (MIT Press, Cambridge, 1991).

John M. Carroll, "What led to the invention," in *The Story of the Laser* (E. P. Dutton, New York, 1964) chapter 2, pp. 60–101.

Sidney S. Charschan, "The evolution of laser machining and welding, with safety," *Proceedings of SPIE Vol. 229* 144–153 (1980).

Russell D. Dupuis, "An introduction to the development of the semiconductor laser," *IEEE Journal of Quantum Electronics QE-23* 651–657 (June 1987).

Raymond C. Elton, *X-Ray Lasers* (Academic Press, Boston, 1990).

J. L. Emmett, W. F. Krupke, and J. I. Davis, "Laser R&D at the Lawrence Livermore National Laboratory for fusion and isotope separation applications," *IEEE Journal of Quantum Electronics QE-20* 591–602 (1984).

T. Y. Fan, "Diode-laser pumped solid-state lasers; historical overview, modeling, wavelength diversity, and comparison to other devices," *IEEE Journal of Quantum Electronics QE-24* 895–912 (1988).

V. I. Grigoryev, *Rem Khokhlov* (Mir Publishers, Moscow, English edition 1985 [Biography of Soviet laser researcher].

Robert N. Hall, "Injection lasers," *IEEE Journal of Quantum Electronics QE-24* 674–678 (June 1987).

Robert N. Hall, "Injection lasers," *IEEE Transactions on Electron Devices, ED-23* 700–704 (1976).

Jeff Hecht, *Beam Weapons: The Next Arms Race* (Plenum Press, New York, 1984), chapter 6 on bomb-driven X-ray laser development.

Jeff Hecht and Dick Teresi, "The short but tempestuous history of the laser," in *Laser: Supertool of the 1980s* (Ticknor & Fields, New York, 1982) chapter 4, pp. 49–61.

Nick Holonyak, Jr., "Early GaAsP lasers development; personal recollections," *IEEE Journal of Quantum Electronics QE-23* 684–691 (1987).

IEEE Journal of Quantum Electronics, special centennial issue on lasers, June 1984.

Gary K. Klauminzer, "Twenty years of commercial lasers—a capsule history," *Laser Focus/Electro-Optics* 54–79 (Dec. 1984).

Willis E. Lamb, Jr., "Laser theory and Doppler effects," *IEEE Journal of Quantum Electronics QE-20* 551–555 (1984).

Willis E. Lamb, Jr., "Physical concepts in the development of the maser and laser," in Behram Kursunoglu and Arnold Perlmutter, ed., *Impact of Basic Research on Technology* (Plenum Press, New York, 1973), pp. 80–85.

Erik Larson, "Patent pending," *Inc.* 105–114 (Mar. 1989) (story of Gould patents).

Lawrence Livermore National Laboratory, *Energy and Technology Review* November 1985, special issue on laboratory X-ray lasers.

Bela A. Lengyel, "Evolution of masers and lasers," *American Journal of Physics 34* 903–913 (1966).

P. Luchini and H. Motz, *Undulators and Free-Electron Lasers* (Oxford Science Publications, Oxford, UK, 1990).

Eliot Marshall, "Gould advances inventor's claim on the laser," *Science 216* 392–395 (Apr. 23, 1982).

Thomas G. Marshall, *Free-Electron Lasers* (Macmillan, New York, 1985).

Allen Maurer, *Lasers: Light Wave of the Future* (Arco, New York, 1982) chapter 3, "Masers to lasers, how they work and the men who made them."

Ivars Melngailis, "Laser development at Lincoln Laboratory," *The Lincoln Laboratory Journal 3,* 3, 347–360 (Fall 1990).

S. Millman, ed., *A History of Engineering and Science in the Bell System: Physical Sciences 1924–1980* (chapter 5, "Quantum electronics—the laser," pp. 151–210.)

Bill Mosely, "Star Warrior Peter Hagelstein" *Omni 11,* 8, 74–78, 91–94 (May 1989) (interview).

M. I. Nathan, "Invention of the injection laser at IBM," *IEEE Journal of Quantum Electronics QE-23* 679–683 (June 1987).

John W. Orton, D. H. Paxman, and J. C. Walling, *The Solid State Maser,* (Pergamon Press, Oxford, 1970).

C. K. N. Patel, "Lasers—their development and applications at AT&T Bell Laboratories," *IEEE Journal of Quantum Electronics QE-20* 561–576 (1984).

A. M. Prokhorov, "Quantum electronics," pp. 110–116 in *Nobel Lectures . . . Physics, 1963–1970* (Elsevier North-Holland, Amsterdam, 1972).

Norman F. Ramsey, "Experiments with separated oscillatory fields and hydrogen masers," *Science 248* 1612–1619 (29 June 1990).

R. H. Rediker, "Research at Lincoln Laboratory leading up to the development of the injection laser in 1962," *IEEE Journal of Quantum Electronics QE-23,* 692–695 (June 1987).

R. H. Rediker, I. Mengailis, and A. Mooradian, "Lasers, their development, and applications at MIT Lincoln Laboratory," *IEEE Journal of Quantum Electronics QE-20* 602–615 (1984).

Arthur L. Schawlow, "From maser to laser," in B. Kursunoglu and A. Perlmutter, ed., *Impact of Basic Research on Technology* (Plenum Press, New York, 1973).

Arthur L. Schawlow, "Masers and lasers," *IEEE Transactions on Electron Devices ED-23* 773–779 (1976).

Arthur L. Schawlow, "Spectroscopy in a new light," (Nobel lecture), *Reviews of Modern Physics 54,* 3, 697–707 (1982).

Arthur L. Schawlow, "Lasers in historical perspective," *IEEE Journal of Quantum Electronics QE-20* 558–561 (1984).

Robert W. Seidel, "From glow to flow: A history of military laser research and development," *Historical Studies in the Physical and Biological Sciences 18* 111–147 (1987).

Robert W. Seidel, "How the military responded to the laser," *Physics Today 41* 12–19 (Oct. 1988).

G. F. Smith, "The early laser years at Hughes Aircraft Co.," *IEEE Journal of Quantum Electronics QE-20* 577–584 (1984).

Howard M. Smith, *Principles of Holography 2nd ed* (John Wiley & Sons, New York, 1975) chapter 1, "Historical introduction," pp. 3–12.

Peter P. Sorokin, "Contributions of IBM to laser science, 1960 to the present," *IEEE Journal of Quantum Electronics QE-20* 585–591 (1984).

A. J. Torsiglieri and W. O. Baker, "The origins of the laser," *Science 199* 1022–1026 (1978).

Charles H. Townes, "Quantum electronics and surprise in the development of technology," *Science 159* 699–703 (1968).

Charles H. Townes, "Production of coherent radiation by atoms and molecules, Nobel Lecture" in *Nobel Lectures . . . Physics 1963–1970* (Elsevier North-Holland, Amsterdam, 1972).

Charles H. Townes, "The early days of laser research," *Laser Focus 14,* 8, 52–58 (Aug. 1978).

Charles H. Townes, "Science, technology, and invention: their progress and interactions," *Proceedings of the National Academy of Sciences USA 80* 7679–7683 (1983).

Charles H. Townes, "Ideas and stumbling blocks in quantum electronics," *IEEE Journal of Quantum Electronics QE-20* 547–550 (1984).

Ronald W. Waynant and Raymond C. Elton, "Review of short-wavelength laser research," *Proceedings of the IEEE 64,* 7, 1059–1092 (July 1976).

Joseph Weber, "Evolution of the laser," *Federation Proceedings, Vol. 24,* Supplement pp. 2–7.

Joseph Weber, ed. *Lasers: Selected Reprints with Editorial Comment Vol. 2* (Gordon and Breach, New York, 1967).

Joseph Weber, ed. *Masers: Selected Reprints with Editorial Comment* (Gordon and Breach, New York, 1967).

An earlier version was published in *Lasers and Optronics* ®
(formerly Lasers and Applications) a Gordon Publications, Inc. publication.

Index

C

D

E

F

G

H

I

J

K

L

M

N

O

P

Q

R

S

T

U

V

W

X

Y

Z